Sustainable Production, Life Cycle Engineering and Management

Series Editors

Christoph Herrmann, Braunschweig, Germany

Sami Kara, Sydney, Australia

SPLCEM publishes authored conference proceedings, contributed volumes and authored monographs that present cutting-edge research information as well as new perspectives on classical fields, while maintaining Springer's high standards of excellence, the content is peer reviewed. This series focuses on the issues and latest developments towards sustainability in production based on life cycle thinking. Modern production enables a high standard of living worldwide through products and services. Global responsibility requires a comprehensive integration of sustainable development fostered by new paradigms, innovative technologies, methods and tools as well as business models. Minimizing material and energy usage, adapting material and energy flows to better fit natural process capacities, and changing consumption behaviour are important aspects of future production. A life cycle perspective and an integrated economic, ecological and social evaluation are essential requirements in management and engineering.

Indexed in Scopus

To submit a proposal or request further information, please use the PDF Proposal Form or contact directly: Petra Jantzen, Applied Sciences Editorial, email:petra.jantzen@springer.com

More information about this series at http://www.springer.com/series/10615

Nadine Madanchi

Model Based Approach for Energy and Resource Efficient Machining Systems

Springer

Nadine Madanchi
Institute of Machine Tools and Production Technology
Technische Universität Braunschweig
Braunschweig, Germany

ISSN 2194-0541 ISSN 2194-055X (electronic)
Sustainable Production, Life Cycle Engineering and Management
ISBN 978-3-030-87539-8 ISBN 978-3-030-87540-4 (eBook)
https://doi.org/10.1007/978-3-030-87540-4

This Springer imprint is published by the registered company Springer Nature Switzerland AG
The registered company address is: Gewerbestrasse 11, 6330 Cham, Switzerland

Acknowledgments

This book has been written in the context of my work as a research engineer at the Chair of Sustainable Manufacturing and Life Cycle Engineering at the Institute of Machine tools and Production Technology (IWF) at the Technische Universität Braunschweig. Herewith, I would like to thank all those who have supported me in the creation of this book.

First of all, I specially thank Prof. Dr.-Ing. Christoph Herrmann as chair and co-director of the institute for providing an excellent working environment with great opportunities and freedom as well as his continuous support over the years. Furthermore, I thank Prof. Dr. Steve Skerlos of the University of Michigan and Prof. Dr. Manbir Sodhi of the University of Rhode Island for their positive feedback and the evaluation of this book as well as Prof. Dr.-Ing. Thomas Vietor of the Technische Universität Braunschweig for the coordination of the dissertation committee. I would further like to thank Prof. Dr. Timothy Gutowski of the Massachusetts Institute of Technology for his great support during my research stay at MIT, which had a strong impact on this work.

Many thanks go to all my former colleagues at the IWF for creating the exceptionally collaborative and positive atmosphere that I could enjoy while working at this institute. Through the intensive time of the last years with exciting business trips, conferences and a lot of time spent at the institute, many of you have also become close friends. In particular, I want to emphasize great gratitude to Dr.-Ing. Marius Winter who already inspired me as a master student for the topics of sustainable machining and cutting fluids and from whose guidance and ideas I have benefited until the successful completion of this book. Further, special thanks go to Prof. Dr.-Ing. Sebastian Thiede the former head of the research group for his insightful comments and reviewing the book. Another great thank you goes to Robar Arafat for his support but also his enthusiasm for the common research topics. Representative for all the students that I have supervised in their theses works or assistant jobs I would like to thank Benjamin Lamottke, Jonas Heinecke, Manuel Mutz, Tobias Rahn, and Wen Tang for their great support and work.

Above all, I am grateful to my parents. Grateful for their encouragement, their unwavering support in all circumstances, and for all the opportunities they have made possible for me. And finally, I would like to thank Roman for joint working sessions in the evenings, weekends, or vacations and especially for everything beyond the dissertation process.

Hamburg, Germany
February 2020

Nadine Madanchi

Contents

Symbols and Abbreviations

Upper Case Symbols

A	Surface area
A_{cm}	Surface area of cleaning tank
A_{cu}	Surface area of uncut chip
A_j	Cutting fluid jet area
A_n	Outlet area of the nozzle
A_{ws}	Surface area of the workpiece
C_i	Constant i
C_r	Radiation coefficient
CI_{cf}	Cutting fluid equivalent costs
CI_e	Energy related equivalent costs
CI_{ml}	Machine tool and labor related costs
CI_{ms}	Machining system related costs
CI_t	Tool related costs
$CO_{2,cf,p}$	Equivalent CO_2 impact due to produced cutting fluid
$CO_{2,cf,d}$	Equivalent CO_2 impact due to disposed cutting fluid
$CO_{2,e}$	Equivalent CO_2 impact due to energy
$CO_{2,t}$	Equivalent CO_2 impact due to tools
D	Milling tool diameter
E	Relative error
E_c	Energy demand due to cutting/processing
E_{cm}	Energy demand due to cleaning
E_{cu}	Energy demand due to cooling
E_{de}	Energy demand due to de-oiling
E_s	Energy demand due to standby stage
E_t	Energy demand due to tool usage
EER	Energy efficiency ratio
EI_{cf}	Cutting fluid equivalent CO_2 impact
EI_e	Energy related equivalent CO_2 impact

EI_{ms}	Machining system related equivalent CO_2 impact
EI_t	Tool related equivalent CO_2 impact
F_{Ad}	Adhesion force
F_c	Cutting force
F_{czm}	Cutting force per grain
F_j	Cutting fluid jet force
F_n	Normal force
Gr	Grashof number
$H_{b,g}$	Hardness of the bulk material of the grain
H_c	Centrifuge height
H_l	Total pressure head loss
H_P	Pump head
$H_{s,g}$	Hardness of the grain surface
H_{sys}	System head
J	Moment of inertia
$K_{c,g}$	Abrasive wear coefficient of the grain
K_{cf}	Correction factor for cutting fluid strategy
$K_{cf,ea}$	Correction factor for cutting fluid drag-out via emissions
K_{cv}	Correction factor for grinding processes
K_m	Correction factor for cutting materials
K_{tw}	Correction factor for tool wear
K_{vc}	Correction factor for cutting speed
K_{vk}	Correction factor for cutting process
K_γ	Correction factor for rake angle
Nu	Nusselt number
P_b	Base load of a machine tool
P_c	Power demand due to cutting power
P_{comp}	Power demand due to compressed air
P_{cu}	Power demand due to cooling
P_d	Power demand due to dressing
P_{de}	Power demand due to de-oiling
P_{ea}	Power demand due to exhaust air fan
$\mathrm{P}_{P,el}$	Electrical power demand due to pump
$\mathrm{P}_{P,m}$	Mechanical power demand due to pump
P_s	Power demand due to spindle
$\dot{Q}_{cf,c}$	Heat input due to cutting process
$\dot{Q}_{cf,con}$	Heat loss due to cutting convection
$\dot{Q}_{cf,cu}$	Heat flow to be removed by cooling unit
$\dot{Q}_{cf,p}$	Heat input due to pumps
$\dot{Q}_{cf,r}$	Heat losses due to radiation
$Q_{cm,i}$	Initial heating load for cleaning fluid
$\dot{Q}_{cm,ev}$	Heat losses due to evaporation during cleaning process
$\dot{Q}_{cm,ev,out}$	Heat losses due to evaporation during workpiece removal
Q_w	Material removal rate
R^2	Coefficient of determination

R_a	Arithmetic mean roughness value
R_c	Centrifuge radius
R_D	Specific gas constant
Re	Reynolds number
T	Temperature
$\bar{T}$	Arithmetic mean of cleaning fluid and air temperature
ΔT	Temperature difference
$\Delta T_{cf,to}$	Tolerable cutting fluid temperature rise
U_d	Dressing overlap ratio
$\dot{V}_{cf}$	Cutting fluid volume flow
$\dot{V}_{cf,P}$	Cutting fluid volume flow at the pump
$V_{cf,total}$	Total cutting fluid volume
V_{cm}	Volume of cleaning medium
$\dot{V}_{ea}$	Exhaust air volume flow
V_n	Amount of compressed air
V_{gt}	Grinding wheel wear volume
V_{gw}	Grinding wheel volume
V_t	Travel volume of cutting fluid
V_{tv}	Minimal tank volume of cutting fluid
V_w	Volume of removed workpiece material
V_{ws}	Size of the workstation volume
Y_i	Observed values
$\bar{Y}_i$	Mean observed values
$\hat{Y}_i$	Modeled values
Z	Number of cutting edges

Lower Case Symbols

a_e	Depth of cut
a_{ea}	Coefficient to determine required air flow
a_{ed}	Dressing depth
a_{ws}	Coefficient to determine wetting
a_p	Width of cut
b	Undeformed chip width
b_{ea}	Coefficient to determine required air flow
b_s	Grinding wheel width
b_{ws}	Coefficient to determine wetting
c	Component
$c_{p,cf}$	Specific heat capacity of cutting fluid
c_{cm}	Specific heat capacity of cleaning fluid
c_{ws}	Coefficient to determine wetting
d_i	Inner pipe diameter
d_g	Diameter of the grinding wheel

d_w	Diameter of the workpiece
e	Pipe element
f	Feed rate
f_{cf}	Interfacial friction coefficient
f_z	Feed rate per cutting edge
h_{cu}	Undeformed chip thickness
$k_{c1.1}$	Specific force to detach a chip with 1 mm^2
k	Constant for determination of tool wear
k_a	Constant for determination of tool wear
k_{ae}	Constant for determination of tool wear
k_f	Constant for determination of tool wear
k_s	Relative wall roughness
k_z	Constant for determination of tool wear
l	Dimensionless distance between adjacent tracks
l_c	Cutting length
l_g	Contact length
l_f	Flow length
l_p	Pipe length
m_c	Mass of chips and scrap material
$\dot{m}_{cf,ea}$	Cutting fluid loss rate due to emissions
$m_{cf,ea}$	Cutting fluid loss due to emissions
$m_{cf,i}$	Initial cutting fluid filling in tank
$\dot{m}_{cf,s}$	Cutting fluid loss rate due to chips wetting
$m_{cf,s}$	Cutting fluid loss due to chips wetting
$\dot{m}_{cf,w}$	Cutting fluid loss rate due to workpiece wetting
$m_{cf,w}$	Cutting fluid loss due to workpiece wetting
$\dot{m}_{cm}$	Evaporated mass rate of the cleaning fluid
m_{de}	Capacity of centrifuge for de-oiling
m_t	Tool loss due to wear
m_w	Mass of workpiece
mc	Constant to determine cutting force
mt	Machine tool
m	Machining system
n	Exponent
n_c	Rotational speed of the workpiece
n_s	Rotational speed of the centrifuge
n_{ws}	Number of produced workpieces
p	Pressure
Δp_i	Local pressure losses
Δp_f	Pressure loss due to pipe friction
Δp_V	Total pressure increase
$p_{cf,l}$	Pressure loss of cutting fluid
p_j	Cutting fluid jet pressure
p_{comp}	Desired pressure level of compressed air
$p_{cf,d}$	Cutting fluid disposal price

$p_{cf,p}$	Cutting fluid procurement price
p_e	Energy procurement price
p_{in}	Initial pressure of compressed air
$p_{l,i}$	Labor rate
$p_{m,i}$	Machine hour rate
p_t	Tool price
$p_{v,cm}$	Fluid vapor pressure of air temperature
r	Evaporation heat at fluid temperature
r_g	Grain tip radius
r_L	Limit radius
Δr_s	Radial grinding wheel wear
r_ε	Corner radius of the cutting tool
s_1	Thickness of the centrifuge shell
s_2	Thickness of the centrifuge bottom
t_c	Processing time
$t_{cf,l}$	Service life of the cutting fluid
t_{cm}	Cleaning time for workpieces
$t_{cm,out}$	Time to remove workpieces from cleaning unit
Δt_{cm}	Difference in temperature
t_d	Machine downtime due to tool and workpiece preparation
t_{de}	De-oiling time
$t_{de,a}$	Acceleration time for de-oiling
t_{in}	Initial temperature of compressed air
t_l	Tool life
t_p	Production time
t_r	Residence time of cutting fluid
t_{tp}	Tool preparation time
t_s	Standby time
t_{wp}	Workpiece preparation time
t_{ws}	Dripping time
v	Flow velocity
v_c	Cutting speed
v_f	Feed speed
v_{fad}	Traverse dressing speed
v_s	Grinding wheel speed
v_w	Workpiece speed
q	Speed ratio
z_c	Number of edges used within a cut

Greek Symbols

α	Attack angle
α^{1-5}	Regression coefficient for emissions during grinding

α_c	Convective heat transfer coefficients
β^{1-5}	Regression coefficient for emissions during turning
β_m	Fluid transfer coefficient
γ	Rake angle of cutting tool
ε_1	Emission level of the surface
η_{CF}	Cutting fluid supply efficiency
η_{comp}	Compressor efficiency
η_{ea}	Exhaust air fan efficiency
η_{hyd}	Hydraulic efficiency
η_{mot}	Motor efficiency
κ	Working cutting edge angle
λ	Friction factor
λ_{air}	Heat conductivity of air
λ_{ke}	Effective grain spacing
μ	Friction coefficient
ν	Kinematic viscosity
ν_{air}	Kinematic viscosity of the air
ν_{cf}	Kinematic viscosity of cutting fluid
ρ_c	Density of centrifuge material
ρ_{cf}	Density of cutting fluid
ρ_{cm}	Density of cleaning fluid
ρ_w	Density of workpiece
ρ_n	Density of gas
σ	Stefan-Boltzmann-constant
σ_{cf}	Surface tension of the cutting fluid
ς	Pressure loss coefficients
φ_c	Cutting direction angle
ω	Angular velocity of centrifuge

Abbreviations

BEP	Best efficiency point
cBN	Cubic boron nitride
$CO_{2\,eq.}$	Carbon dioxide equivalents
EU	European Union
EER	Energy efficiency ratio
GWP	Global warming potential
FEM	Finite element method
KEA	Cumulated energy demand
LCA	Life cycle assessment
LCC	Life cycle costing
MFCA	Material flow cost accounting
MFO	Multifunctional oil

MQL	Minimal quantity lubrication
OEL	Occupational exposure limit
OP	Operating point
RQL	Reduced quantity lubrication
SEC	Specific energy consumption
SME	Small and medium-sized enterprises
USA	United States of America
VB	Flank wear
VBA	Visual basic for applications
VDI	Verein Deutscher Ingenieure
VFD	Variable frequency drive

List of Figures

List of Tables

Chapter 1
Introduction

This chapter aims at the motivation behind the work by presenting the current context and challenges for energy and resource efficiency of machining systems (Sect. 1.1). Following this, the research objectives are derived and the outline of the book is presented (Sect. 1.2).

1.1 Motivation and Scope

The economic growth of the last century was accompanied by an increasing consumption of energy and resources. While initially highly industrialized countries contributed most to the global consumption of natural resources, today emerging and developing countries play an increasing important role. With the rapidly growing population and improvement in living standards, the demand for energy and resources will continue to rise (Fischer-Kowalski et al. 2011). By its very nature, the continuation of this development will at some point reach planetary boundaries that will limit growth prospects and poses a threat to human well-being (OECD 2011). This raises the demand for a sustainable development "that meets the needs of the present without compromising the ability of future generations to meet their own needs" (WCED 1987).

In light of this development, the EU increased its climate ambition and presented the *European Green Deal.* The vision of the Green Deal is to achieve climate neutrality by 2050 with no net emissions of greenhouse gases. The EU has already reduced the greenhouse gas emissions by 23% between 1990 and 2018, despite an economic growth of 61% (EU 2019). Nevertheless, the EU still generated almost 4 billion tonnes of CO_2 equivalents by industry and households in 2018 (see Fig. 1.1). The figure also indicates that about one quarter of the industry induced emissions are caused by manufacturing. This accounts for almost 0.8 Gt CO_2 equivalents (Eurostat 2020). Thus, manufacturing is particularly relevant in terms of a sustainable development.

N. Madanchi, *Model Based Approach for Energy and Resource Efficient Machining Systems*, Sustainable Production, Life Cycle Engineering and Management,
https://doi.org/10.1007/978-3-030-87540-4_1

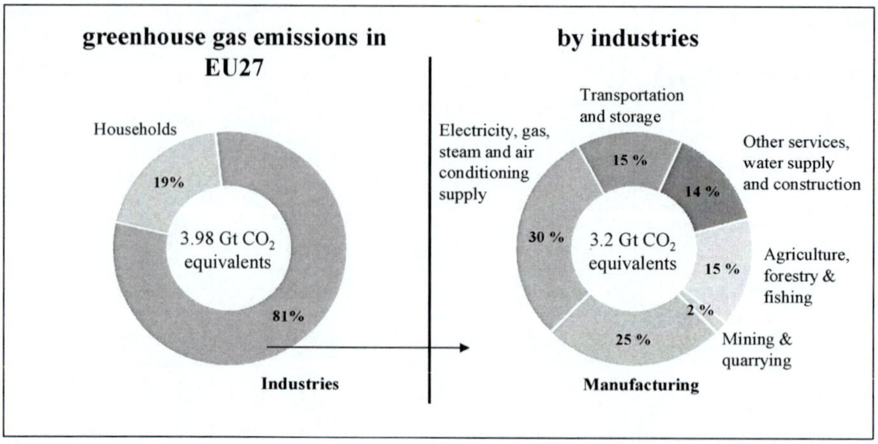

Fig. 1.1 Greenhouse gas emissions by industry and househols in the EU27[1] in 2018 [based on Eurostat (2020)]

In general, manufacturing represents an important economic force and a major mechanism regarding economic development, which has a significant impact on the environment and society (Gutowski et al. 2013). Sustainable manufacturing can be defined as "the creation of manufactured products through economically-sound processes that minimize negative environmental impacts while conserving energy and natural resources. Sustainable manufacturing also enhances employee, community and product safety" (EPA 2017). Past studies have estimated an energy saving potential of up to 20% for the manufacturing industry by 2050 (Gutowski et al. 2013).

Within the manufacturing industry, the metalworking sector accounts for around 10% of the total manufacturing value added in the EU27[2] and comprises nearly a fifth of the total number of companies in manufacturing with a significant share of employees (Ecorys Consulting 2009). For production purposes about 3.5 million metalworking machine tools were operated throughout Europe. These include in particular machine tools for processes using geometrically defined cutting edges, i.e. milling and turning processes (Schischke et al. 2012a). Grinding processes, in contrast, are among the processes using geometrically undefined cutting edges and although grinding machines account for a smaller share, they are still particularly relevant. In industrial countries, it is estimated that about 20 to 25% of the expenditure for all machining processes result from grinding (Malkin and Guo 2008).

The operation of machine tools requires electrical energy. For Europe the required energy is in the range of 200–300 TWh per year (Schischke et al. 2012b). According to Abele et al. (2011) "a modern medium-sized machine tool consumes approximately 12 times more electrical energy than a one-family house". However, Dahmus and

[1] EU27 represents 27 member states of the European Union considered without United Kingdom.

[2] EU27 represents 27 member states of the European Union with United Kingdom, but without Croatia.

Gutowski (2004) demonstrated that only less than 15% of the energy consumption is caused by the cutting energy during machining. Based on the exemplary energy demand of a machine tool shown in Fig. 1.2, it can be noted that the peripheral subsystems are main consumers and account for a large share of the energy. Especially cutting fluid systems, which supply filtered fluid to the machining process are relevant and also have an impact on exhaust air systems, which clean the air from aerosols and particles. Cutting fluid systems furthermore affect the need for subsequent subsystems such as the de-oiling of chips and cleaning of workpieces, which also consume energy.

Further research studies have demonstrated that the cutting fluid induced energy consumption is on average between 25 and 30% of the overall consumption of the machining system (Li et al. 2011; Triebs 2015). The research by Bode (2007) even indicated a total energy consumption of more than 50% due to supply and preparation of coolants. Besides the demand of energy, the use of cutting fluids is also connected with a large consumption of resources. Only in Germany, the annual use of cutting fluids is estimated at 791,000 t per year with the majority being based on mineral oil (Herrmann et al. 2017). Due to this large consumption of cutting fluids and also the need for additional peripheral systems, cutting fluids account for around 7 to 17% of the workpiece-related manufacturing costs (Klocke and Eisenblätter 1997). Nevertheless, the common use of cutting fluids is based on the fulfilment of technological tasks. Their primary purpose is to cool and to lubricate the contact zone between tool and workpiece to ensure high workpiece quality and to decrease tool wear (Klocke 2009). Cutting fluids therefore have a significant impact from an environmental and economic as well as technological perspective. The specific impact, however, depends on the particular cutting fluid strategy, which can be differentiated by the fluid formulation and applied amount.

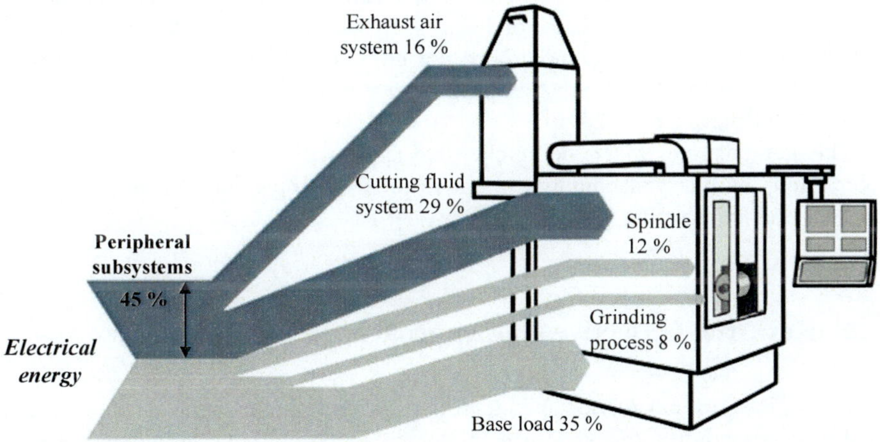

Fig. 1.2 Exemplary energy distribution of a grinding machine [machine tool from Posselt (2016)]

This generally indicates the relevance of machining systems in connection with cutting fluid strategies for a more sustainable development and thus the need for action. Although the need to integrate sustainable development in the machine tool industry is obvious, it still poses a challenge to implement it in concrete measures. In general, the following three strategies were derived as possible approaches for it (Dyckhoff and Souren 2008; Herrmann 2010). The *efficiency strategy* - with the goal of minimizing energy and resource consumption over the life cycle with maximal output, the *sufficiency strategy* - focusing on behavioral change towards an adequate resource usage and the *consistency strategy* - implementing the use of natural products and renewable energy by creating a new system, process or product. Especially the last strategy has a great potential for improvement, but also requires great effort to be implemented and is therefore designed for the long term. In practice, the implementation of the efficiency strategy is strongly preferred by companies, as environmental savings due to reduced usage of energy and materials often also result in economic savings. For the implementation of specific measures, it is essential to consider all interdependencies in order to avoid problem shifting (Thiede 2012).

With respect to machining systems, the application of the efficiency strategy would therefore imply a complete elimination or significant reduction of cutting fluid usage. This, however, would be an improper simplification, due to the technological importance of cutting fluids, which also needs to be taken into account. Otherwise this could lead to problem shifting due to, for example, an increased compressed air demand for chip removal or fluid supply (Madanchi et al. 2015; Priarone et al. 2016) or an increased tool wear and negative impacts on the workpiece quality (Madanchi et al. 2019). Besides the amount of cutting fluids, there is further potential to increase the efficiency by adapting the design of peripheral systems to the machining process and to improve the process parameters. In general, Duflou et al. (2012) have demonstrated that based on literature an improved control and setting of process parameters can increase the energy efficiency with a factor of 1.1 and a redesign of machine tools and selective control even by a factor of 1.2 to 3. However, regarding these approaches the possible increase of other input flows e.g. compressed air or tools has to be taken into account as well as the effect on productivity and workpiece quality (Madanchi et al. 2019; Priarone et al. 2016).

This demonstrates that a comprehensive approach is needed that involves the interactions within a machining system in order to apply the efficiency strategy without problem shifting.

1.2 Research Objective and Approach

In view of the current situation and identified challenges, machining systems have to be designed not only with regard to technological functions, but also from an environmental and economic perspective. As the use of cutting fluids has a great impact on all three perspectives, it is the general objective of this work to develop a comprehensive approach supporting the development and implementation of measures to increase the energy and resource efficiency of machining systems in relation to cutting fluid strategies.

To achieve this objective the following aspects have to be considered:

- Identification of relevant peripheral subsystems and components as well as their interactions with the machining process. Milling, turning and grinding systems have been identified as being particularly relevant. The focus is on single processes as well as process chains including multiple processes.
- Development of specific models for the machining system elements that consider the effect of process parameters, cutting fluids, process chains and further machinery characteristics on the energy and resource efficiency.
- Assessment of alternative designs, parameters and strategies from an environmental, economic and technological perspective.
- Development of improvement approaches in order to increase the energy and resource efficiency of machining systems.

To address the above-mentioned aspects, the work is divided into six chapters. The structure including chapter-specific objectives is visualized in Fig. 1.3 and briefly described below.

Following this introduction, Chap. 2 presents the required technical background in context of machining systems with a focus on turning, milling and grinding as well as relevant peripheral subsystems. This also includes the functions and strategies of cutting fluids. Further, methodical approaches are presented to convert a system into simplified models as well as to assess the performance of the system. This state of art serves as the foundation of the work.

Chapter 3 synthesizes and analyzes the state of research regarding machining systems. In that context and based on the previous results, criteria are derived to select relevant research approaches as well as criteria to evaluate the degree to which relevant aspects are fulfilled. The research gap revealed by this comparison demonstrates the need for further research.

Chapter 4 proposes a concept to increase the energy and resource efficiency of machining systems in relation to the use of cutting fluids. It involves the interactions of system elements by modeling their relevant input and output flows. To ensure model quality, these are mainly validated by experimental results. The assessment of the interactions from a technological, environmental and economic perspective as well as measures to improve it, are also part of this chapter. The chapter closes with the application cycle of the concept and prototypical implementation into a software environment.

Chapter 5 demonstrates the application and potential of the developed concept based on different case studies. These case studies focus on the assessment and improvement of existing machining systems as well as the planning of a new process chain with multiple machining process. The application also includes the exemplary implementation of organizational and technical improvement measures.

The work closes in Chap. 6 with a summary, critical review of the results and outlook on further research in this field.

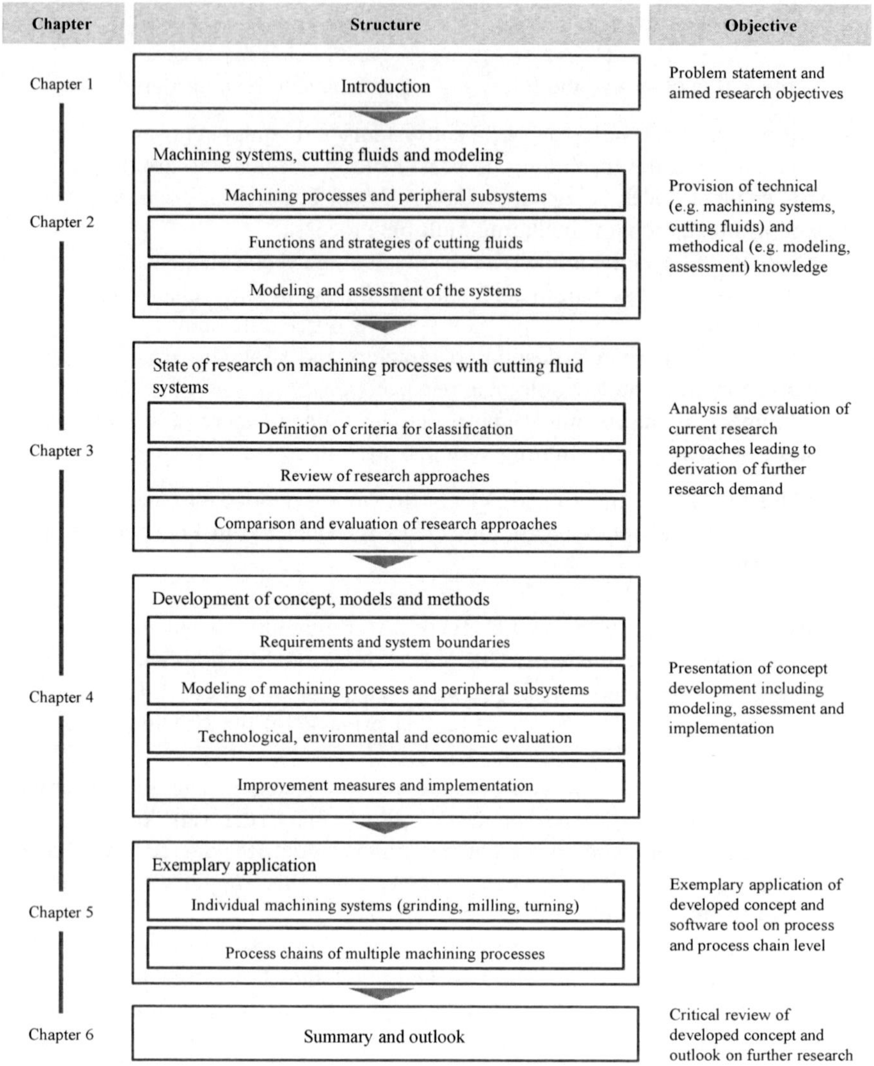

Fig. 1.3 Structure and objectives of the work

References

Abele E, Kuhrke B, Rothenbücher S (2011) Energieeffizienz spanender Werkzeugmaschinen. Wissenschaftsmagazin der TU Darmstadt, Germany

Bode H-O (2007) Einfluss einer energieeffizienten Produktion auf Planungs- und Produktprämissen am Beispiel der Motorenfertigung. XII. Internationales Produktionstechnisches Kolloquium, pp 299–305

Dahmus JB, Gutowski TG (2004) An environmental analysis of machining. In: Proceedings of ASME International Mechanical Engineering Congress and R&D Exposition, pp 643–652. https://doi.org/10.1115/IMECE2004-62600

Duflou JR, Sutherland JW, Dornfeld D, Herrmann C, Jeswiet J, Kara S, Hauschild M, Kellens K (2012) Towards energy and resource efficient manufacturing: a processes and systems approach. CIRP Ann 61:587–609. https://doi.org/10.1016/j.cirp.2012.05.002

Dyckhoff H, Souren R (2008) Nachhaltige Unternehmensführung: Grundzüge industriellen Umweltmanagements. Springer, Berlin, Germany

Ecorys Consulting (2009) Competitiveness of the EU metalworking and metal articles industries: FWC sector competitiveness studies

EPA (2017) Sustainable manufacturing. https://www.epa.gov/sustainability/sustainable-manufacturing#:~:text=Sustainable%20manufacturing%20is%20the%20creation,employee%2C%20community%20and%20product%20safety. Accessed 11 Sept 2020

EU (2019) The European green deal: communication from the commission. https://eur-lex.europa.eu/legal-content/EN/ALL/?uri=CELEX:52019DC0640. Accessed 11 Sept 2020

Eurostat (2020) Greenhouse gas emission statistics—air emissions accounts: [env_ac_ainah_r2] (statistical data). https://ec.europa.eu/eurostat/statistics-explained/index.php?title=Greenhouse_gas_emission_statistics_-_air_emissions_accounts&oldid=471850#Developments_between_2008_and_2018. Accessed 10 Sept 2020

Fischer-Kowalski M, Swilling M, Weizsäcker EU von, Ren Y, Moriguchi Y, Crane W, Krausmann F, Eisenmenger N, Giljum S, Hennicke P, Romero Lankao P, Siriban Manalang A, Sewerin S (2011) Decoupling natural resource use and environmental impacts from economic growth. A report of the Working Group on Decoupling to the International. United Nations Environment Programme

Gutowski TG, Allwood JM, Herrmann C, Sahni S (2013) A global assessment of manufacturing: economic development, energy use, carbon emissions, and the potential for energy efficiency and materials recycling. Annu Rev Environ Resour 38:81–106. https://doi.org/10.1146/annurev-environ-041112-110510

Herrmann C (2010) Ganzheitliches Life Cycle Management - Nachhaltigkeit und Lebenszyklusorientierung in Unternehmen. VDI-Buch. Springer, Berlin, Germany

Herrmann C, Madanchi N, Winter M, Öhlschläger G, Greßmann A, Zettl E, Schwengers K, Lange U (2017) Ökologische und ökonomische Bewertung des Ressourcenaufwands - Wassermischbare Kühlschmierstoffe. Technical Report, VDI Zentrum Ressourceneffizienz GmbH (VDI ZRE)

Klocke F, Eisenblätter G (1997) Dry cutting. CIRP Ann 46:519–526. https://doi.org/10.1016/S0007-8506(07)60877-4

Klocke F (2009) Manufacturing processes 2: grinding, honing, lapping. Springer, Berlin, Heidelberg, Germany

Li W, Zein A, Kara S, Herrmann C (2011) An investigation into fixed energy consumption of machine tools. In: 18th CIRP international conference on life cycle engineering, Braunschweig, Germany, pp 268–273. https://doi.org/10.1007/978-3-642-19692-8_47

Madanchi N, Kurle D, Winter M, Thiede S, Herrmann C (2015) Energy efficient process chain: the impact of cutting fluid strategies. In: 22th CIRP international conference on life cycle engineering, Sydney, Australia, vol 29, pp 360–365. https://doi.org/10.1016/j.procir.2015.02.056

Madanchi N, Thiede S, Gutowski TG, Herrmann C (2019) Modeling the impact of cutting fluid strategies on environmentally conscious machining systems. In: 26th CIRP international conference on life cycle engineering, vol 80, West Lafayette, pp 150–155. https://doi.org/10.1016/j.procir.2019.01.068

Malkin S, Guo C (2008) Grinding technology: theory and applications of machining with abrasives, 2nd edn. Industrial Press, New York

OECD (2011) Green growth studies: energy. OECD Publishing

Posselt G (2016) Towards energy transparent factories. Dr.-Ing. dissertation, Technische Universität Braunschweig. Springer, Switzerland

Priarone PC, Robiglio M, Settineri L, Tebaldo V (2016) Modelling of specific energy requirements in machining as a function of tool and lubricoolant usage. CIRP Ann 65:25–28. https://doi.org/10.1016/j.cirp.2016.04.108

Schischke K, Hohwieler E, Feitscher R, König J, Kreuschner S, Wilpert P, Nissen NF (2012a) Energy-using product group analysis—lot 5: machine tools and related machinery. Task 2 Report—Economic and Market Analysis, Berlin, Germany

Schischke K, Hohwieler E, Feitscher R, König J, Kreuschner S, Wilpert P, Nissen NF (2012b) Energy-using product group analysis—lot 5: machine tools and related machinery. Task 1 Report—Definition, Berlin, Germany

Thiede S (2012) Energy efficiency in manufacturing systems. Dr.-Ing. dissertation, Technische Universität Braunschweig. Springer, Berlin, Germany

Triebs JB (2015) Analyse, Optimierung und Simulation des Energieumsatzes dezentraler Fluidkreisläufe von Werkzeugmaschinen. Dr.-Ing. dissertation, RWTH Aachen. Apprimus Verlag, Aachen, Germany

WCED (1987) Our common future: the Brundtland-report. World Commission on Environment and Development. Oxford University Press, Oxford

Chapter 2
Machining Systems and Impact of Cutting Fluids

This chapter presents the theoretical basis of this work. First, the levels of abstractions in a machining system are presented as well as relevant interactions between them (Sect. 2.1). As a decisive influencing factor on design and operation of the system, cutting fluids are then presented with regards to functions and strategies (Sect. 2.2). Following this, the subsystems are described in more detail regarding task and energy and resource consumption (Sect. 2.3) as well as their interactions within machining process chains (Sect. 2.4). Approaches for modeling the described systems and impacts are described in the following (Sect. 2.5). The chapter concludes with a preliminary statement on the potential need for further research (Sect. 2.6).

2.1 Elements and Interactions in Machining Systems

Machining systems are within the focus of this work. These systems consist of various subsystems and components as presented in Fig. 2.1.

From a systems engineering perspective, systems are divided into three hierarchical levels. At the lowest level are the components, which can be described as "a basic functional unit of a system" (Misra 2008). The combination of different components to form another functional unit can be considered as a subsystem. It performs a specific sub-task, but usually requires other subsystems or systems. A combination of subsystems, however, to perform a more complex and defined task, is described as a system (Misra 2008). The division between the three levels thus depends on the perspective and the overall task to be performed. In the context of machining systems, tools or pipes are, for example, considered as components, while the exhaust air system with the subtask of air filtration is regarded as a subsystem.

The relevant subsystems and their interactions are presented in Fig. 2.2. The figure indicates that the considered machining system requires various energy and resource input flows such as unfinished workpieces and tools, which in return result in

N. Madanchi, *Model Based Approach for Energy and Resource Efficient Machining Systems*, Sustainable Production, Life Cycle Engineering and Management,
https://doi.org/10.1007/978-3-030-87540-4_2

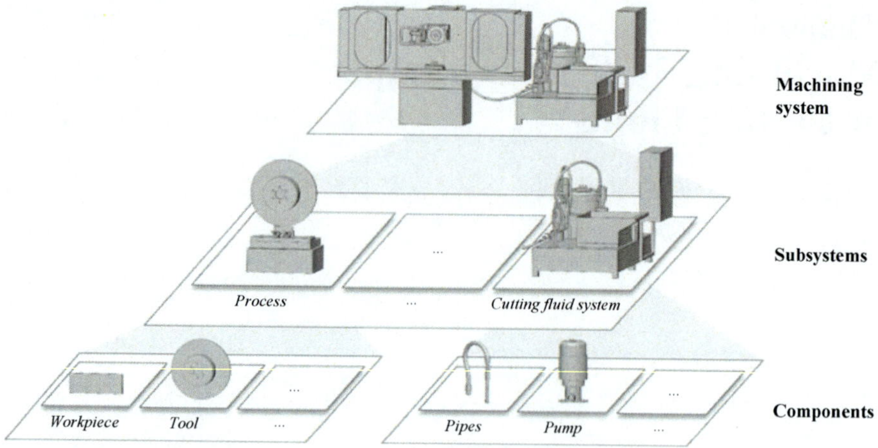

Fig. 2.1 Levels of abstractions in a machining system

output flows such as finished workpieces and worn tools. According to the efficiency strategy presented in the previous chapter, these flows should be targeted and adapted to foster a sustainable development. The efficiency of the system can be derived from its definition as the "relationship between the result achieved and the resources used" (DIN EN ISO 9000 2015). In this context, the finished workpieces could be considered as the achieved result and input flows such as electrical energy, tools, cutting fluid etc. as used resources. According to the efficiency strategy, reducing the input flows e.g. amount of cutting fluid, number of used tools or alternatively increasing the achieved results e.g. number of finished workpieces, are approaches towards sustainability.

In view of the considered machining system, the actual *machining process* is at the center of the consideration. It produces the technically desired and value-adding changes to the workpiece. In order to fulfil this task successfully, various factors such as the selected process parameters, the tools but also the type of cutting fluid strategy are important. Cutting fluid strategies differ in terms of fluid composition and applied amount. The strategies therefore include not only flooding but also dry machining, as presented in more detail in Sect. 2.2. The choice of cutting fluid strategy is not only important regarding the machining process itself, it also has a direct impact on the peripheral subsystems (Madanchi et al. 2019c). *Cutting fluid systems* are designed as a circulation system, which consists of further subsystems to return the fluid and chips from the machine tool, to filter and to prepare the cutting fluid as well as to supply it back to the machining process. The chips and scrap material removed from the machined workpiece are then further processed (VDI 3035 2008). With regard to the scrap processing, the cutting fluid strategy also influences the need for a *de-oiling* process such as pressing or centrifuging to remove the fluid from the

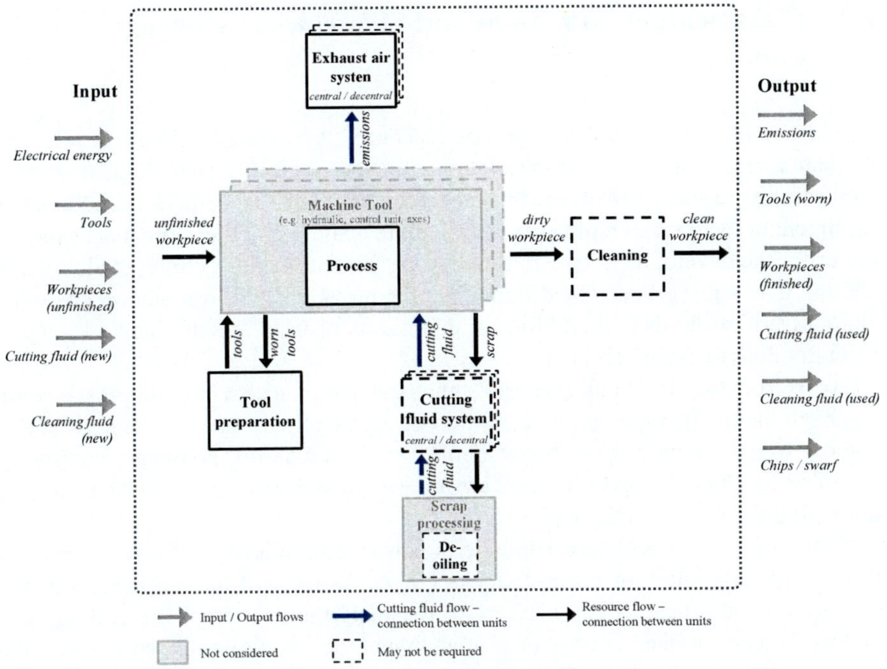

Fig. 2.2 Interactions between relevant subsystems [based on Madanchi et al. (2019c)]

chips and swarf. The *cleaning* of the workpiece might be required as an intermediate cleaning at different stages through the machining process chain or at the final stage in preparation for e.g. assembling and coating (Mang and Dresel 2007). Moreover, the generation of emissions to air during the machining process is also influenced by the use of cutting fluids. Thus, the installation of an *exhaust air system* to remove and to filter the emissions from the machine tool is required. However, also in the case of dry machining, the use of an exhaust air system is recommended (VDI 3802-2 2012). The generation of emissions also depends strongly on the setting of the machining process parameters. Similarly, the tool wear is also affected by multiple factors such as process parameters and cutting fluids. The subsystem *tool preparation* comprises both, the replacement of worn tools or inserts as well as the dressing of grinding wheels in the case of grinding machines.

Against this background, the scope of this work is limited to systems and subsystems that are directly influenced by cutting fluid strategies. Consequently, the *machine tool* as a whole or the *scrap processing* will not be discussed in detail, as they are hardly directly influenced. In Sect. 2.3, the relevant subsystems as well as their relevant components are presented in more detail to further consider the interactions within the entire machining system during strategy implementation.

2.2 Classification, Functions and Strategies of Cutting Fluids

The impact of cutting fluid strategies on machining systems is particularly relevant. As demonstrated in Fig. 2.3, these strategies are used to fulfill technological aspects, while aiming to satisfy environmental and economic ones as well. The simultaneous fulfilment of the three perspectives is difficult to achieve and therefore makes them a driver for the development of different cutting fluid strategies (Winter et al. 2012a). Cutting fluid strategies can be differentiated in terms of fluid formulation involving alternative base fluids and additives and in terms of used amount of fluid ranging from flooding to dry machining.

Apart from these two characteristics, the strategies can also be distinguished with regard to the media type and its condition of aggregation. Further strategies in this context would be the use of, for example, solid lubrication, cryogenic cooling or supercritical CO_2. However, as these are not yet commonly used in practice, they are not within the focus of this work.

Cutting fluids are widely used in the metalworking industry to fulfil the primary tasks of cooling and lubricating (Klocke 2009). As most of the mechanical power is converted into heat, it is necessary to dissipate the heat from the cutting zone and workpiece surface to prevent a change in material properties. Due to lubrication and thus reduction of friction between workpiece and tool, the heat generation as well as tool wear and energy demand can also be reduced. Further, the lubricating effect also prevents roughness peaks from welding against each other on moving surfaces, which improves the surface quality (Klocke 2009; VDI 3397-1 2007). In addition to these primary functions, cutting fluids need to fulfill secondary tasks as well. These include flushing and removal of chips and solids from the machining

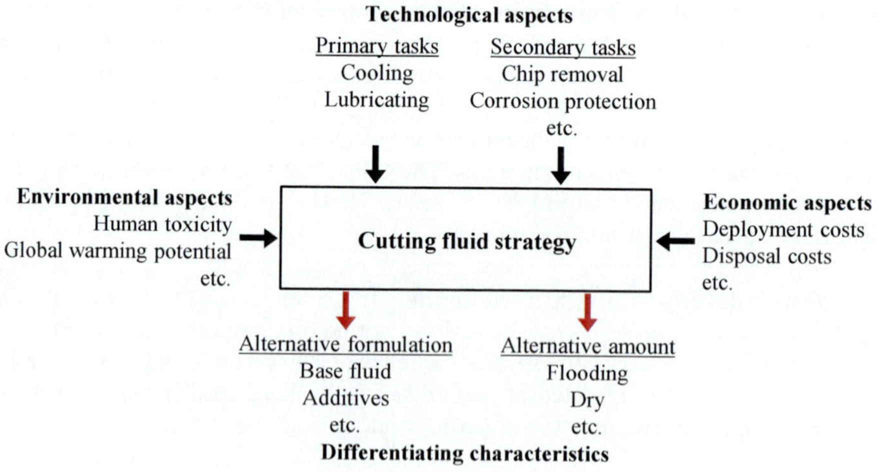

Fig. 2.3 Drivers and characteristics of cutting fluid strategies

location, corrosion protection of both machine tool and workpiece as well as in case of grinding also cleaning of the grinding wheel (Klocke 2009). The fulfilment of these tasks should enable an efficient machining process with increasing material removal rates, extended tool life and improved surface quality (VDI 3397-1 2007). The use of cutting fluids is therefore essential for various processes and in particular for processes with high demands on accuracy and surface quality such as grinding. The extensive use, which was estimated at 791,000 t in 2014 (Herrmann et al. 2017), also places additional requirements on cutting fluids. On the one hand, cutting fluids should be resistant to aging by being chemically and physically stable and on the other hand, they should also be biodegradable in order to be disposed later, while they must not affect the environment and health of machine operators (Denkena and Tönshoff 2011). However, the fulfillment of the mentioned tasks and requirements strongly depends on type and composition of the fluid.

According to DIN 51385, cutting fluids are divided into oil-based and water-based cutting fluids (DIN 51385 2013). The classification of cutting fluids is also presented in Fig. 2.4.

The oil-based fluids are not water-miscible and usually based on mineral oil, but also natural and synthetic ester oils, synthetic hydrocarbons or polyglycols can be used (VDI 3397-1 2007). While these base fluids are used to define initial properties of the cutting fluid, additional chemical substances (additives) are used to specifically influence the fluid properties to achieve the desired functions and may account for up to 25% by weight (Brinksmeier et al. 2015). Without these additives an efficient machining process is limited or not even possible (Möller and Nassar 2002).

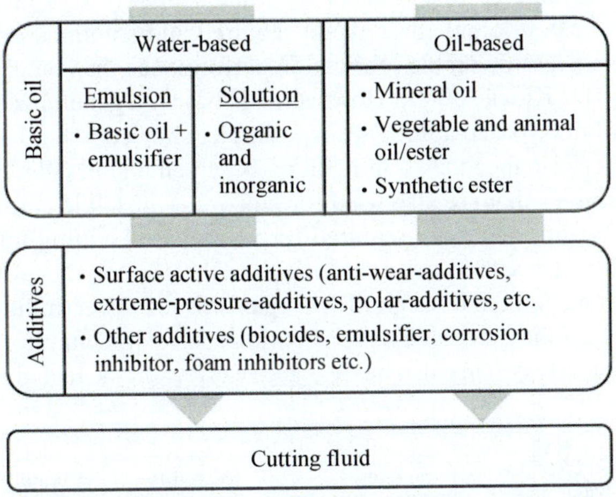

Fig. 2.4 Classification of cutting fluids [Madanchi et al. (2019b), based on Brinksmeier et al. (1999), Winter et al. (2012b)]

Generally, additives can be classified into the two main groups of fluid stabilizing and tribologically active substances. The fluid stabilizing substances, which change the physical and chemical properties of the cutting fluid, include additives such as oxidation and foam inhibitors. The tribologically active additives (also described as anti-wear additives), which change the friction conditions, include additives such as polar additives, extreme pressure additives, and anti-wear additives (Brinksmeier et al. 2015; Klocke 2009). In comparison to water-based cutting fluids oils have an advantageous lubricating performance, which is mainly caused by the higher viscosity of oil compared to water in connection with the used additives. However, a disadvantage of the oil is the relatively poor cooling effect, due to the lower heat capacity and thermal conductivity (Klocke 2009). The use of non-water-miscible cutting fluids also leads to an increased generation of oil mist and oil vapors, which can be explosive and must be removed by an exhaust air system for health and safety reasons (DGUV 2012). Table 2.1 compares the thermal-physical characteristics of water and mineral oil.

In contrast to non-water-miscible cutting fluids, water-based cutting fluids are mixed with water before use and can further be distinguished between emulsions or solutions (DIN 51385 2013). An emulsion consists of fine oil droplets, which are dispersed in water. In this case emulsifiers are used to ensure that the fluid remains stable and that the two phases do not separate (Möller and Nassar 2002). For machining processes with defined cutting edges concentrations of 4–15% and for processes with undefined cutting edges such as grinding concentration of 3–5% are commonly used (VDI 3397-1 2007). Mineral oils, natural and synthetic ester oils or polyglycols can, for example, be used as base oils (VDI 3397-1 2007). Solutions are mostly free of mineral oil and the concentrate is dissolved or finely dispersed in water. Thus, the use of emulsifiers is not required and solutions provide a greater stability but also a less favorable lubrication performance than emulsions (Denkena and Tönshoff 2011; Klocke 2009). However, this is only a generalization. The development of a polymer based solution, for example, showed a similar or even better technological performance than a mineral oil based emulsion (Winter et al. 2012a). Besides the additives that are also used for non-water-miscible cutting fluids, additional stabilizing substances are required for water-based cutting fluids. Due to the high water content, these fluids provide conditions that support the growth of microorganisms such as bacteria and fungi. Microorganisms, however, influence the cutting fluid's condition, as they metabolize, for example, corrosion inhibitors or emulsifiers (Rabenstein et al. 2009). This also has a negative effect on the technical performance

Table 2.1 Thermal-physical characteristics of water and mineral oil (Denkena and Tönshoff 2011)

Parameter	Variable	Water	Mineral oil
Spec. heat capacity	c_p [J/(g K)]	4.2	1.95
Density	ρ_{cf} [g/cm^3]	0.99	0.85
Viscosity at 40 °C	ν_{cf} [mm^2/s]	0.66	9.5

and may lead to increased tool wear and surface roughness (Koch 2007). For this reason, biocides are commonly used as additives to eliminate microorganisms and to reduce the microbial contamination. Further additives such as corrosion and foaming inhibitors are also particularly important for water-based cutting fluids (Möller and Nassar 2002).

Besides conventional cutting fluids, multifunctional oils (MFO) can also be used. These can be applied for at least two different applications such as for the hydraulic system and machining process or multiple machining processes (VDI 3397-1 2007). Using such MFOs can prevent problems caused by leakages, which can, for example, result in contamination of emulsions and thus increased maintenance effort, shorter service life and production downtimes (Möller and Nassar 2002).

As demonstrates in Fig. 2.3, cutting fluid strategies cannot only be differentiated with regard to the aforementioned formulation of the cutting fluid, but also to the applied amount of cutting fluid. Each strategy has a specific impact on the lubricating and cooling performance. As presented in Table 2.2, the strategies flooding, reduced quantity lubrication (RQL), minimal quantity lubrication (MQL) as well as dry machining are usually distinguished by the amount of applied cutting fluid.

According to Table 2.2, amounts of up to 100 l/min and more are used for the flooding strategy. In general, the decision whether to use oil- or water-based cutting fluids for flooding depends on the individual machining process and requirements. While oil-based cutting fluids are commonly used for machining processes that require a high lubricating effect, water-based cutting fluids are used for processes with high heat generation due to their cooling effect (Brinksmeier et al. 2015). However, for a variety of different machining processes, both oil- or water-based cutting fluids can be used and the decision often depends on experiences (Brinksmeier et al. 2015) or on peripheral parameters such as the design of the machine tool or process chain (VDI 3397-1 2007). The use of solutions, however, is mainly limited to grinding processes and in general they are not used for processes with defined cutting edges (Klocke 2011). Globally, the share of water-based cutting fluids, especially emulsions, is more than 80% (Meister 2011), with earlier literature even referring to 90% (Zwingmann 1979). This is also reflected in the distribution of disposed cutting fluids in Germany in 2014 (Herrmann et al. 2017). This may also be due to more complex protection systems that are required for machine tools in order to use oil-based cutting

Table 2.2 Classification of cutting fluid strategies [according to Dietrich (2016), VDI 3397-1 (2007)]

Type of fluid application	Strategy description	Quantity of applied fluid	System design
Full fluid application	Flooding	5 to >100 l/min	Circulating system
Reduced fluid application	Reduced quantity lub	50 ml/h to 2 l/h	Circulating system
	Minimal quantity lub	<50 ml/h	Loss lubrication
No application	dry	No fluid	No system

fluids (Meister 2011). Due to the need for such additional peripheral equipment and the large amount of cutting fluid, conventional flooding also poses a considerable cost factor with 7–17% of the workpiece-related manufacturing costs (Klocke and Eisenblätter 1997). Moreover, skin diseases due to contact with cutting fluids have been considered as one of the most common occupational health diseases in the metalworking industry and also the inhalation of cutting fluid emissions can lead to serious health problems (Suuronen 2009).

Despite the technical advantages of the flooding strategy, such problems have resulted in the development of alternative strategies, which significantly reduce or even eliminate the amount of used cutting fluid. With flow rates of less than 2 l/h for RQL or even less than 50 ml/h for MQL, both strategies represent possible alternatives. Usually, non-water-miscible fluids are applied and especially synthetic esters are recommended for MQL machining (Suda et al. 2002; Weinert et al. 2004). In this case, compressed air and small amounts of cutting fluid are used and sprayed as a fine air-oil mixture by an internal or external supply into the contact zone. Since the cutting fluid completely disperses in the form of emissions, MQL can also be described as loss lubrication (Mang and Dresel 2007). As a result, the environmental as well as economic impact related to the procurement, filtration and disposal of cutting fluids can be reduced. However, the use of small cutting fluid quantities also leads to a reduced chip removal and heat dissipation from the contact zone, which is particularly critical for machining processes with geometrically undefined cutting edges such as grinding. The MQL strategy is, therefore, not suitable for those machining processes with high demands regarding geometric accuracy and surface quality (Mang and Dresel 2007; VDI 3397-1 2007). Dry machining, in contrast, even completely eliminates the use of cutting fluids. This absence of tribologically effective fluids lead to more friction between the tool and the workpiece, which results in higher temperatures and thus increased tool wear (Klocke and Eisenblätter 1997). Geometrical accuracies and surface quality of the workpiece, chip removal as well as production rates were identified as further challenges due to dry machining. Possible approaches to meet these challenges focus on the choice of tool materials, tool coatings or tool geometry. So far dry machining has only been implemented for certain materials of good machinability and machining processes with moderate cutting parameters (Goindi and Sarkar 2017).

Overall, it is required to consider the effects of each strategy not only on the machining process itself, but on the entire machining system including the peripheral subsystems as well as on the achievable cycle times. Depending on the process setting and parameters, the use of MQL could also lead to an increased environmental impact due to the compressed air demand (Priarone et al. 2016) or dry machining could result in an increased economic impact due to the increased tool wear (Madanchi et al. 2019c).

Apart from distinguishing cutting fluid strategies based on the used amount of fluid, strategies can also be distinguished with regard to the condition of aggregation and type of media. Further strategies in this context would be the use of, for example, solid lubrication, cryogenic cooling or supercritical CO_2. However, as these are not yet widely used in practice, they are not within the focus of this work.

2.3 Description of Relevant Elements of the Machining System

According to Fig. 2.2, the different subsystems as well as relevant components of the considered machining system are presented in more detail in the following.

2.3.1 *Machining Processes and Tool Preparation*

Cutting is one of the main groups of manufacturing processes and can further be divided into subgroups such as machining processes with geometrically defined and undefined cutting edges (DIN 8580 2020). In case of all machining processes a certain layer of workpiece material is removed by relative motions between cutting tool and workpiece. The motion consists of a primary motion and a feed motion, resulting in a continuous or repeated removal of chips during a rotation (Boothroyd and Knight 2006).

2.3.1.1 Machining with Geometrically Defined Cutting Edges

According to (DIN 8589-0 2003), machining processes with geometrically defined cutting edges are distinguished. These machining processes can further be divided into groups using single-point tools such as turning or multipoint tools such as milling (Boothroyd and Knight 2006). As previously mentioned, this work is limited to turning and milling processes, but other processes such as drilling or sawing are also performed with defined cutting edges. In case of turning, the relative motion is usually performed by a rotating movement of the workpiece and a feed motion performed by the cutting tool. Turning is therefore applied to produce rotationally symmetric workpieces. In contrast to this, milling is characterized by a rotating movement performed by the cutting tool and a feed motion, which is vertical or oblique to it. While the primary motion is always performed by the cutting tool, the feed motion can be performed by the tool or the workpiece. In contrast to turning, milling is performed with an interrupted cut without continuous contact of the cutting edge, due to the use of multipoint tools (Denkena and Tönshoff 2011). Generally, it can be distinguished between a roughing and finishing process. While roughing focuses on a high material removal rate, finishing focuses on a high surface quality.

In dependence on the different processes and workpiece materials, a variety of different cutting tools can be used. They can generally be classified into tool steels such as high speed steels, cemented carbide, ceramics and super-hard cutting tool materials such as boron nitride or diamond. The choice of material is particularly critical, as it has a major influence on the tool wear and therefore tool consumption and tool changing times. For this reason, research constantly focuses on the development and improvement of materials as well as coatings for the tools. Especially cemented carbides are coated with a thin layer of hard materials such as nitrides, carbon nitrides and/or oxides in order to improve wear resistance and performance (Klocke 2011). The properties of coatings are also particularly important for machining processes without the conventional use of cutting fluids. Here, the coatings have to compensate for the lubricating properties of the cutting fluids (Sales et al. 2017). The choice of cutting tool and tool material is also connected with a specific embodied energy. It represents the total amount of energy required to produce the tool including the raw materials, transportation and disposal. The tool selection is therefore also critical in terms of the environmental impact (Dröder et al. 2020; Kirsch et al. 2014).

Apart from tool material and coating, the tool wear depends on other factors such as the process parameters or the workpiece material. Depending on the material combination and process setting, different wear mechanisms can occur. Abrasion is a mechanical wear, which is caused by sliding of hard particles between the active partners. In the case of adhesive wear, small particles are micro-welded by atomic bonding forces. Furthermore, tribochemical wear results from diffusion or oxidation during machining. In general, these wear mechanisms strongly depend on the process temperature and lead to a deformation of the tool (Denkena and Tönshoff 2011). Common wear forms of a turning tool are presented in Fig. 2.5. It can generally be distinguished between wear notches, craters and flank wear. A common tool wear criteria is the width of flank wear land (VB) of around 0.3–0.6 mm (ISO 3685 1977).

A special machining process with geometrically defined cutting edges is the hard machining of materials with a hardness above 47 HRC (Denkena and Tönshoff 2011). In some cases, it can substitute grinding, as it achieves a high surface quality and accuracy of hardened workpieces. Depending on the workpiece geometry it is for example possible to reduce various process steps by hard turning instead of grinding. However, hard machining is still less common as it leads to high tool wear and requires a machine tool with high stiffness. The development and improvement of tool materials and coatings therefore enables a wider use (Denkena and Tönshoff 2011; Klocke 2011).

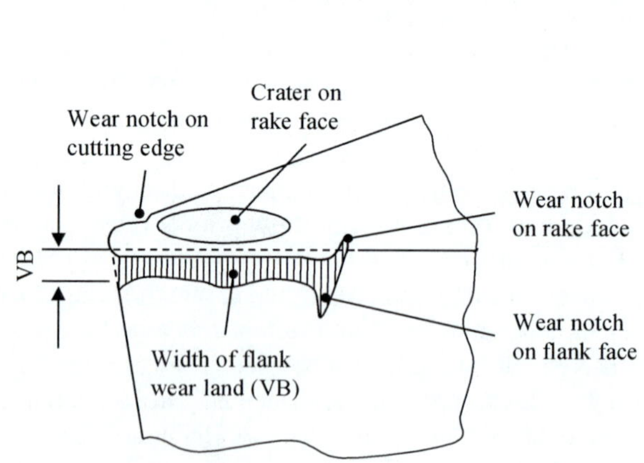

Fig. 2.5 Wear forms of turning tools [according to Klocke (2011)]

2.3.1.2 Machining with Geometrically Undefined Cutting Edges

According to DIN 8589-0 (2003), different machining processes with geometrically undefined cutting edges are classified. This work is limited to grinding processes, but honing or lapping are other examples for processes using undefined cutting edges. Based on the classification grinding processes differ significantly from turning and milling processes. They are characterized by using abrasive wheels and are usually used at the end of a process chain for finishing purposes. The achievable average surface roughness (R_z) of the workpiece is between 0.2 and 10 μm for grinding, while is it between 2.5 to 160 μm for milling and between 1 to 250 μm for turning (Klocke 2011). Besides the high surface quality, grinding also enables high dimensional accuracy of the workpiece. As visualized in Fig. 2.6 a grinding wheel consists of abrasive grains that are the cutting edges, bond material for linking and positioning of the grains as well as pores to transport chips and cutting fluid (Rowe 2009). Thus, the type of bond and grain material, size of the grit and porosity define the properties and performance of these tools.

Regarding conventional grinding wheels, aluminum oxide or silicon carbide are commonly used as abrasive grains with vitrified or resinoid bonds. Abrasive grains from diamond or cubic boron nitride are used for superabrasive wheels with vitrified, resin or metal bonds (Malkin and Guo 2008). Similar to the tools for turning or milling, the material of the grinding wheels is also decisive with regard to the environmental impact (Kirsch et al. 2014; Winter et al. 2015). Although the embodied energy of a grinding wheel might be significantly higher compared to an end mill, the usability for processing and tool life of a grinding wheel is usually also several times higher (Kirsch et al. 2014).

When selecting the grinding wheel for a specific process, the wear mechanisms should also be considered. High pressure and temperature result in grit and bond

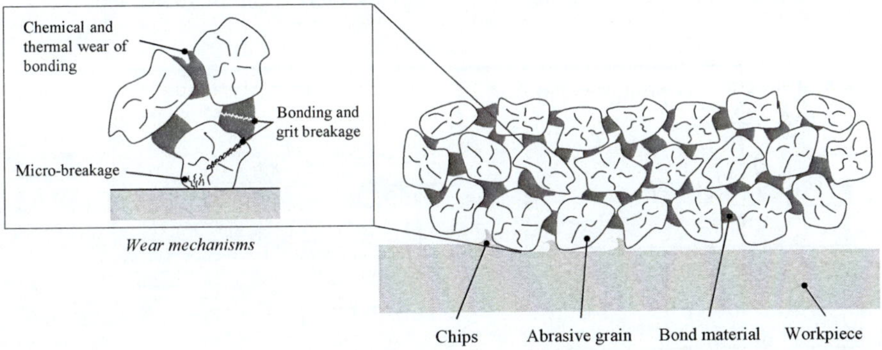

Fig. 2.6 Structure and wear forms of grinding wheels [based on Brecher and Weck (2019), Klocke (2009)]

wear (see Fig. 2.6). Thus, the abrasion resistance of the grit material is reduced by oxidation and diffusion processes at the contact zone and leads to a dulling of the abrasive grit. Furthermore, thermal and mechanical influences lead to grain and bond breakage, which results in the detachment of abrasive particles and grains from the bond. Additional chemical influences can also lead to bond wear (Klocke 2009; Malkin and Guo 2008). Due to these wear forms, a dressing process is required in preparation for the actual grinding process. The dressing process aims at ensuring accuracy, achieving sharp cutting edges, removing the bond around the abrasive grains and cleaning the wheel (Rowe 2009).

2.3.2 *Cutting Fluid System*

The cutting fluid system is a circulation system that includes the subsystems of cutting fluid supply to transport the fluid to the machine tool, cutting fluid return back from the machine tool as well as the cutting fluid preparation to maintain and ensure the condition of the fluid (VDI 3035 2008). The cutting fluid system is influenced by the choice of cutting fluid strategy, as it is particularly required for flooding. Its design also depends on the applied fluid amount and fluid properties such as viscosity or density. The subsystems and their relevant components are presented in the following.

2.3.2.1 Cutting Fluid Supply

In order to supply the required amount of cutting fluid into the contact zone, further components such as a pump, piping system and nozzle are necessary. A configuration of these components is illustrated in Fig. 2.7. By selecting and dimensioning each of these components appropriately, the energy efficiency can be increased.

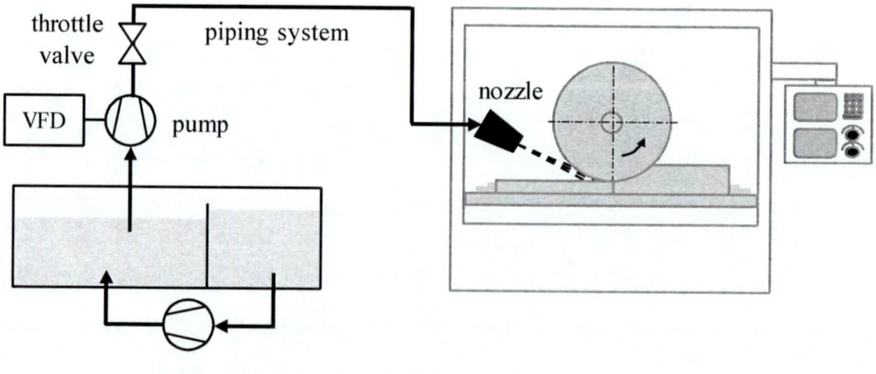

Fig. 2.7 Components for the cutting fluid supply

Pump

Pumps are required for the actual delivery of cutting fluid. They are used to transport the fluid from the tank to the machine tool, but also to transport it to the filtration unit within the tank (see Fig. 2.7). Depending on the specific requirements centrifugal pumps, screw pumps or gear pumps are best suited to deliver cutting fluids. The selection can be made based on the required pressure, where an internal supply through a tool requires a higher pressure level compared to external supply, but which requires high volume flows (Triebs 2015). Other selection criteria also consider the viscosity or share of solids in the cutting fluid (VDI 3035 2008). An overview of commonly used cutting fluid pumps is presented in Table 2.3.

In practice, screw pumps are used for the high pressure supply of cutting fluid, for example, for the internal cooling of cutting tools or cleaning of grinding wheels (Helpertz et al. 2010; Kuhrke 2011). Centrifugal pumps, by contrast, are commonly used for conventional external cutting fluid supply or machine bed flushing (Triebs 2015). Overall, these are also the most commonly used type of pumps due to the low price, simple handling and maintenance effort (DOE 2014). The selection and operation of the pump are particularly critical, as it is associated with high energy consumption. Salonitis and Ball (2013) determined that only the cutting fluid pump accounts for 45% of the total energy demand of a cylindrical grinding process. This is also relevant from an economical perspective, as energy costs make up 45% of the total costs over the lifecycle of a pump (Brecher et al. 2012).

In order to increase the energy efficiency, it is important to consider the entire pump system. Pumps are driven by electric motors where the efficiency can be essentially increased by two approaches. One approach focuses on the use of a variable frequency drive (VFD) to adjust the required motor speed and thus to achieve the desired volume flow of the cutting fluid. With this approach, the use of a throttle can be replaced to control the flow. The impact of using a VFD instead of a throttle is exemplarily demonstrated in Fig. 2.8. Within this example, the energy demand can be reduced by around 30%.

The effect of using a VFD can also be observed with regard to performance curves. Each pump is characterized by a specific curve, which indicates the pump head in relation to the volume flow. The operating point (OP) is located at the intersection of the performance curve with the system curve and should be close to the best efficiency point (BEP). This is demonstrated on the example of a centrifugal pump in Fig. 2.9. It further presents the impact of using a throttle valve in comparison to

Table 2.3 Overview of exemplary pumps [with permission according to VDI (3035 2008)]

Type of pump	Pressure	Fields of application
Centrifugal pump (single-stage)	Up to 10 bar	All low-viscosity media
Centrifugal pump (multi-stage)	Up to 50 bar	All media
Screw pump	Up to 140 bar	Oils and lubricative media
Gear pump	Up to 200 bar	Oils and lubricative media with low solids content

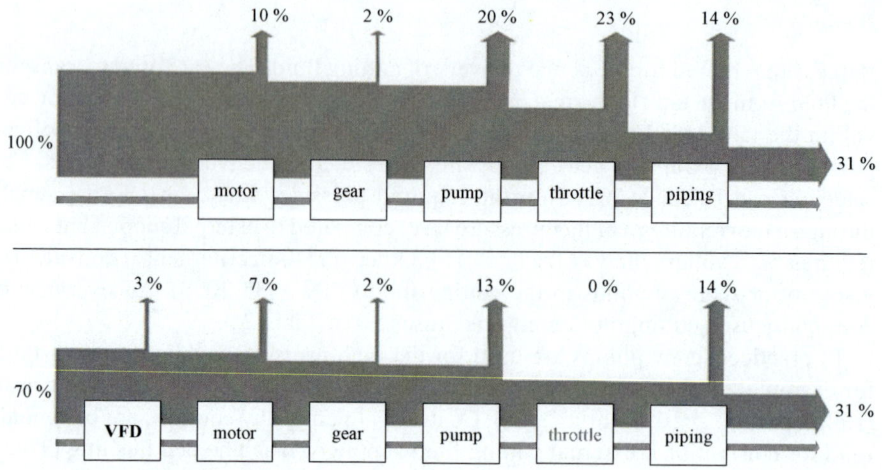

Fig. 2.8 Sankey diagram of exemplary energy dissipation within a pump system - without VFD (top), with VFD (bottom) [based on LfU (2009)]

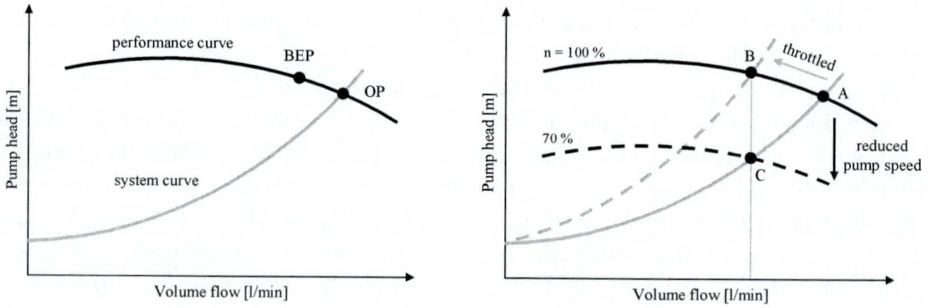

Fig. 2.9 Pump performance curve (left) and regulation approaches (right) [based on DOE (2014), KSB (2005)]

an adjustable speed drive to control the volume flow. In order to achieve the desired volume flow, closing the throttle increases the pressure and takes the operating point from *A* to *B*. Using a speed control, in contrast, changes the pump performance curve by reducing the motor frequency. Thus, operating point *B* and *C* provide the same desired volume flow of cutting fluid, but with a reduced pressure and power demand at point *C* compared to point *B* (DOE 2014; KSB 2005).

There are similar effects with screw pumps, but in this case it is not the pump performance curve but the volume flow that is adjusted by a using a VFD. This can lead to energy savings up to 70%. In addition to the direct energy savings, this approach also leads to a reduced heat input into the fluid, pump wear as well as noise development (Helpertz et al. 2010).

Besides using a speed control, another approach to increase the efficiency focuses on the appropriate selection of a high efficiency motor that meets the power demand

and does not have to be further regulated. In cases where no significant variations of the cutting fluid volume flow are needed, such as for the machine bed flushing, this approach might be recommended as the use of a VFD is also connected to energy dissipation (Hesselbach 2012). The motors are mainly three-phase asynchronous motors, which are classified according to energy efficiency classes from IE1 to IE4 (DIN EN 60034-30-1 2014). As of 2017, only electric motors with the energy efficiency class of IE3 or higher may be distributed within the European Union. Motors of the energy efficiency class IE2 can still be offered with variable speed control for the power range from 0.75 to 7.5 kW (Hesselbach 2012). By using a 0.75 kW motor with the energy efficiency class IE4 instead of IE1, the efficiency can for example be increased by more than 10% (DIN EN 60034-30-1 2014).

Piping system

A piping system is required in order to deliver the cutting fluid from the tank to the nozzle. Cutting fluid that flows through the piping system creates friction on the inner walls and thus leads to a pressure loss. Besides friction, the overall pressure losses are also caused by local losses in valves or fittings. To compensate the pressure loss and to guarantee the required cutting fluid supply, additional pump power must be provided. It is indicated that the piping system may account for an energy dissipation of 14% within the supply system (see Fig. 2.8). An appropriate design of the piping system for the cutting fluid supply could, therefore, contribute to a reduced energy demand (Wittmann 2007).

With regard to the pressure loss due to friction, the friction factor depends on the relative roughness as well as the Reynolds number. These correlations are based on empirical studies and can be summarized within the Moody chart. While the Reynolds number is critical to calculate the friction factor in case of a laminar flow, the relative roughness is critical in case of a turbulent flow (Wittel et al. 2017). In order to determine the roughness of the inner wall, it is required to consider the type and condition of the used pipe. In practice, the use of welded or flanged steel pipes is recommended (Astakhov and Joksch 2012). Further pressure loss is caused by additional components such as valves and fittings or expansions and contractions within the piping system. In literature a dimensionless loss coefficient is assigned to each of these elements.

However, for both types of pressure loss, the flow velocity is particularly important. In this case the venturi-effect has to be considered, which describes that the flow velocity is inversely proportional to a change in the flow cross-section. The continuity equation further states that the volume flow of incompressible fluids is everywhere constant in a pipe (Sigloch 2008). High flow velocities thus result in small pipe diameters, but also in high pressure losses. For an economical flow velocity, there are recommendations for cutting fluids based on practical experiences. While the flow velocity of an emulsion should be between 1.0 to 1.5 m/s for a suction pipe and 2.0 to 2.5 m/s for a pressure pipe, the flow velocity of an oil should be 0.5–1.0 m/s for a suction pipe and 1.0–2.0 m/s for a pressure pipe (Triesch 2010).

Nozzle

The nozzle links the cutting fluid system to the machining process. As design and type of the nozzle have a significant impact on the cutting fluid speed and flow, the selection of a suitable nozzle is important to increase the amount of cutting fluid delivered through the contact zone between tool and workpiece. Thus, by selecting a suitable nozzle for a particular process, it is possible to reduce the pump energy and improve the efficiency (Heinzel 2016; Madanchi et al. 2017; Seidel et al. 2020). An appropriate fluid delivery is particularly relevant for grinding, as an air barrier circulates the grinding wheel and has to be overcome by the cutting fluid (Marinescu et al. 2004). A large variety of different nozzles are used in practice and can be distinguished according to function, focus and geometry of the nozzle (Heinzel 1999; Marinescu et al. 2007). A respective classification of cutting fluid nozzles is presented in Table 2.4.

The application of flooding nozzles such as free jet and shoe nozzles are most common for grinding processes (Brinksmeier et al. 1999). Conventional free jet nozzles have a rather simple design that is independent from the profile and dimension of the tool, which allows them to be used flexible for different machining applications. However, due to flooding and thus a large oversupply of cutting fluid, the share of useful cutting fluid delivered through the contact zone is rather small (Marinescu et al. 2007). Shoe nozzles are particularly used for grinding and positioned as close to the grinding wheel as possible to remove the air barrier. This results in a longer contact between cutting fluid and grinding wheel, which enables the grinding wheel to

Table 2.4 Classification of cutting fluid nozzles [based on Brinksmeier et al. (1999), Marinescu et al. (2007)]

Function	*Geometry*	*Nozzle*	*Volume flow*	*Position*
Flooding	Slot nozzle Round / pipe nozzle	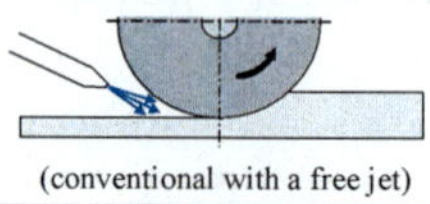 (conventional with a free jet)	↑↑	Tangential for conventional free jet nozzles
	Shoe nozzle	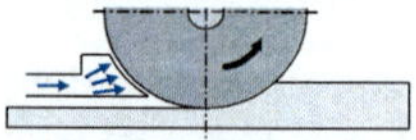	↑	Close to grinding wheel
Not flooding	Needle nozzle		→	Directed to contact zone
	Spot jet / point nozzle	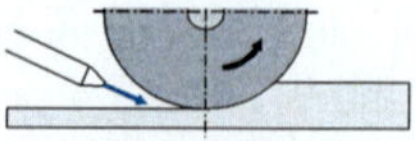	→	Tangential
	Spray nozzle	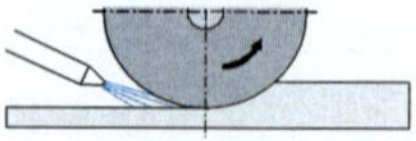	↓↓	Close to contact zone

Volumne flow: very high ↑↑ high ↑ medium → low ↓ very low ↓↓

accelerate the fluid up to wheel speed and improves the wetting of the wheel surface. In contrast to free jet nozzles, their geometrical inflexibility is a disadvantage (Klocke 2009; Marinescu et al. 2004). A non-flooding nozzle is the spot jet or point nozzle that is characterized by a small outlet cross section and high fluid velocities, which facilitate to overcome the air barrier (Brinksmeier et al. 1999). The combination of multiple point nozzles is also described as a needle nozzle (Marinescu et al. 2007). Here, different tubes can be adjusted depending on the grinding wheel or workpiece profile (Klocke 2009). Spray nozzles are usually used for MQL and require only small amounts of cutting fluid. While these nozzles can be used for turning and milling, they are of limited use for grinding processes, as the air barrier is critical for these nozzles and the cooling performance is insufficient (Denkena and Tönshoff 2011).

Besides the type of nozzle, the design and position are also essential. In order to create a coherent cutting fluid jet with minimal entrapped air, several design principles have to be considered. A sharp-edged nozzle outlet and the avoidance of sudden cross-sectional changes with transitions or bevels greater than 7° are recommended. Further, a high surface quality inside the nozzle and a minimal straight nozzle length are advised (Brinksmeier and Bohling 2008). A common nozzle design with concave inner walls is named Rouse nozzle and was first used by Webster et al., as a cutting fluid nozzle (Webster et al. 1995). Regarding the positioning of jet free nozzles the distance and angle to the contact zone are relevant. Various studies have already investigated the influence of positioning for grinding processes. It was concluded that the nozzle should be close to the grinding wheel (Brinksmeier and Bohling 2008; Engineer et al. 1992; Gviniashvili 2003) and tangential with an angle of 10° to the wheel to improve the cooling performance (Brinksmeier and Bohling 2008).

The presented nozzles are used to deliver cutting fluid from the outside to the tool, but it is also possible to deliver cutting fluid through the tool. In this case, the machine spindle, tool holder and tool have to be specially designed and adapted to this strategy. The internal cutting fluid supply is usually used for processes with geometrically defined cutting edges to increase tool life, break chips and/or assure dimensional accuracy. A disadvantage of this strategy is the low volume flow of cutting fluid, which may not be sufficient to remove the chips from the machining area (Astakhov and Joksch 2012). Also, with regards to grinding a cutting fluid supply through the grinding wheel has been investigated (Graham and Whiston 1978), but it is not commonly used in practice.

In addition to the nozzles that are used to deliver cutting fluid into the contact zone, further nozzles are used for different tasks. Flushing nozzles are generally used to remove chips from the machining area of the machine tool, while additional cleaning nozzles are used specifically for grinding processes to flush particles from the wheel surface (Klocke 2009; Triebs 2015).

2.3.2.2 Cutting Fluid Return and Chip Transportation

The used cutting fluid as well as the produced chips have to be removed from the machine tool into a preparation system. Depending on the cutting fluid strategy, type of chips as well as constructional condition of the machine tool alternative variants can be used. The different concepts (1–5) are illustrated in Fig. 2.10.

Concept 1: Within this concept cutting fluid and chips are transported to a cleaning unit though a piping system and without the use of additional pumps. It is also described as gravity return flow. This is an energy efficient possibility, but it requires a suitable pipe gradient of at least 1–2%. If the gradient is not sufficient, additional flushing nozzles can be used to support the transportation. The flushing pressure has to be about 2.5 bar, which increases the cutting fluid volume by 30% up to 100% (Astakhov and Joksch 2012; VDI 3035 2008). As the flushing function generally operates unregulated, the energetic benefit of this variant is then reduced (Rahäuser 2015). Therefore, this concept is particularly recommended for a centralized cutting fluid system that is located in the basement below the machine tools or still has all constructional options during factory planning (Astakhov and Joksch 2012). The performance of the concept increases with larger chip sizes.

Concept 2 and 3: These concepts are characterized by the use of return pump stations. They deliver the chips containing cutting fluid coming from the machine tool via a closed piping system to the cleaning unit. In the second concept, short chips are also pumped, while the third concept integrates an additional chip treatment step for longer chips. Within this step, they can be separated from the cutting fluid or shredded to a pumpable size by chip breakers (VDI 3035 2008). These return pump stations are more flexible than gravity return flow systems and usually used when

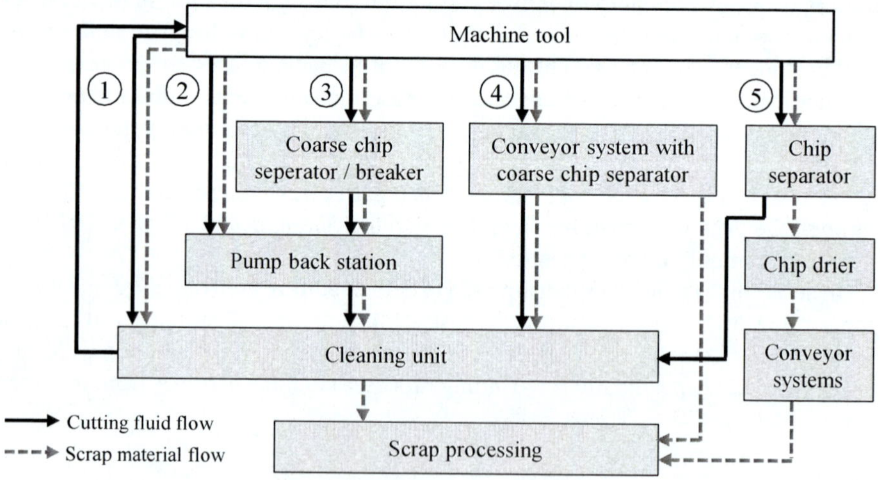

Fig. 2.10 Components for the cutting fluid return and chip transportation - alternative concepts (with permission from VDI 3035 2008)

machine tools and cutting fluid systems are located on the same level. However, due to the additional pump the mechanical load on the fluid as well as the energy demand are increased (Astakhov and Joksch 2012).

Concept 4: For this concept, cutting fluid containing short as well as long chips is already treated in the return flow in order to separate coarse chips. A collection tank is often used for this purpose, in which the chips sediment and are then removed with a conveyor system, e.g. a scraper conveyor or slat-band conveyor. Subsequently, the cutting fluid with the short chips is delivered to the cleaning unit, while the coarse and long chips are transported to the chip processing unit (Triesch 2010; VDI 3035 2008). This concept is rather relevant for machining processes with geometrical defined cutting edges as these processes generate longer chips.

Concept 5: Within this concept the cutting fluid is separated from most chips and returned to the cleaning unit. In contrast to concept 4, the reduced long chips are removed together with the short chips by a pneumatic conveying system and transported to the chip processing (VDI 3035 2008).

2.3.2.3 Cutting Fluid Preparation

In order for the cutting fluid to fulfil its functions properly, it must be prepared accordingly. Therefore, a filtration unit is required to remove chips and dirt from the fluid and a cooling unit may be used to actively cool the cutting fluid. Further, it is also important that a sufficient amount of fluid is available in the tank. Relevant components are presented in Fig. 2.11.

Cooling unit

Heat input into the cutting fluid occurs through different sources. During standby state it is especially the heat input of the filtration and circulation pump that influences the temperature of the cutting fluid. A specific case study demonstrated that about 24% of the electrical power consumption of a pump is liberated as heat into the cutting fluid

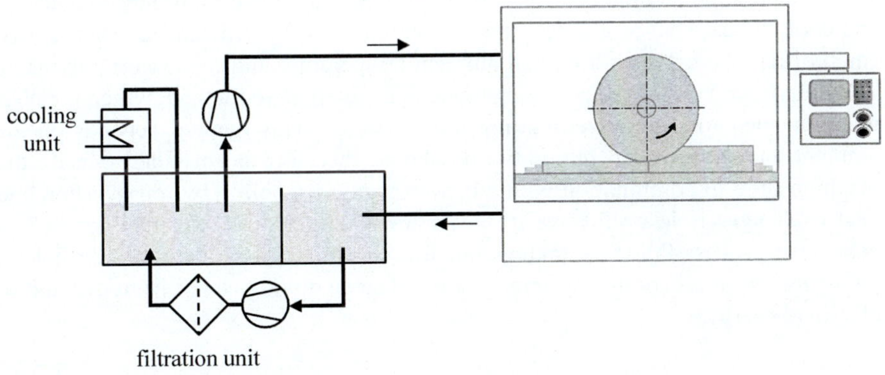

Fig. 2.11 Components for the cutting fluid preparation

(Herrmann et al. 2009). During the processing state there is additional heat from the supply pump as well as heat from the machining process itself. Most of the mechanical cutting energy from the cutting process is also converted into heat and the ratio is considered to be greater than 90% (Döbbeler et al. 2015; Klocke 2011). In the contact zone of an exemplary turning process, the total heat flow is distributed between the tool with 18%, the workpiece with 7% and the chips with 75%. However, this distribution strongly depends on the thermal conduction coefficients of the materials and used process parameters. Higher cutting velocities lead to higher temperatures, but also to a significantly lower proportion entering the tool (<1%) (Klocke 2011; Müller 2004). By using cutting fluid, 66–87% of the overall heat flow resulting from the machining process enters into the cutting fluid (Knobloch 1979). Besides these possibilities of heat input into the cutting fluid, there are also heat losses due to cutting fluid evaporation and convection into the ambience.

A rise of the cutting fluid temperature can be critical for reasons of geometrical and dimensional accuracy and should therefore be controlled. Especially regarding water-based cutting fluids, a temperature above 30 °C leads to ideal growth conditions for microorganisms and problems with skin contact (VDI 3397-2 2007). The dimensioning of the cutting fluid tank, the piping and the amount of circulating cutting fluid are therefore crucial for controlling the temperature. As the temperature of the cutting fluid is usually higher than the ambient temperature, heat is dissipated from the fluid to the environment through the pipes and tank walls (Kuhrke 2011). In order to improve this heat dissipation, the design of a flat tank with a large surface area should be preferred. The residence time of the cutting fluid inside the tank should be 6–10 min for water-based fluids and 10–20 min for oils in order to achieve cooling by natural convection (VDI 3035 2008).

However, in case these design measures are not sufficient to keep the cutting fluid at the desired temperature level, the fluid has to be actively cooled using a cooling unit. Generally it is aimed that the cutting fluid temperature is plus or minus two to three degrees of the ambient temperature (Byers 2006). For this purpose, different cooling systems can be used, although the decision often depends on the available space, type of cutting fluid, climate or energy cost (Astakhov and Joksch 2012). A possible method is the use of air coolers that blow air through a condenser and cool by heat convection. However, this method is not efficient for the required small difference between fluid and ambient temperature and therefore rarely used. The evaporation system of a cooling tower is an alternative method, which is rather recommended for centralized cutting fluid systems. This method depends on the ambient temperature and humidity and should therefore be installed outside the machine shop in combination with a heat exchanger. Cooling by refrigeration and heat exchanger is less sensitive to the ambient temperature and may be used to heat other process fluids (Astakhov and Joksch 2012; Byers 2006). Nevertheless, the operation of all cooling systems is also connected with a significant demand of electrical energy.

Filtration unit

Used cutting fluids are predominantly contaminated by chips, tool wear residues and tramp oil. This contamination could negatively affect the process results and therefore cutting fluid has to be continuously cleaned and prepared. Generally, three different configurations of the filter and cleaning units can be distinguished: full-flow filtration, filtration on by-pass and by-pass/partial flow filtration (VDI 3397-2 2007). These configurations are demonstrated in Fig. 2.12.

In case of a full-flow filtration, the unit is placed in direct line between machine tool and cutting fluid tank. This also means that process stability is compromised, if the filtration unit fails, unless an alternative solution is provided (VDI 3397-2 2007). The most common application of this alternative is for decentralized cutting fluid systems (Astakhov and Joksch 2012). The use of by-pass cleaning systems has its advantage regarding low investment and operating costs, but results in a higher level of fluid contamination as the majority of solids will only be separated after several cleaning cycles (Mang and Dresel 2007; VDI 3397-2 2007). By-pass systems are commonly used in addition to a full-flow filtration (Astakhov and Joksch 2012). In case of a partial flow system, a separate filtration loop with a second fluid tank and pump is used. Despite the high energy consumption of the pump, partial flow filtration is the most commonly used configuration in practice as it achieves minimal contamination of the fluid (Astakhov and Joksch 2012; Mang and Dresel 2007; VDI 3397-2 2007).

For the actual cleaning of the cutting fluid different methods are used in practice. These can be distinguished between separation systems, which use physical mechanisms such as gravity or magnetization and filtration systems, which introduce an additional media or filter into the cutting fluid (Byers 2006). Further literature also makes a general distinction according to whether they operate with or without the use of consumables (Reik and Wnuk 2014). As the actual filtration process is not considered to be a main driver of energy and resource consumption, it is not discussed in more detail for the purpose of this work.

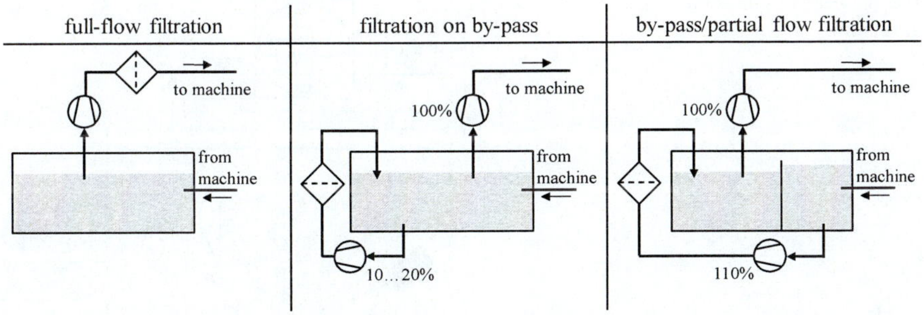

Fig. 2.12 Alternative configurations of cutting fluid filtration units [with permission according to VDI 3397-2 (2007)]

2.3.3 Exhaust Air System

The use of cutting fluids in machining processes leads to the formation of emissions. This is particularly critical with regard to occupational health and safety. Depending on the size of the particles and thus where they deposit in the respiratory tract, the inhalation of the emissions can be hazardous to health. Further, a certain concentration of oil emissions within the machine tool can also become explosive (DGUV 2012). For this reason, modern machine tools are not only enclosed, but also equipped with an exhaust air system to remove and filter the emissions as shown in Fig. 2.13. The reduction and control of cutting fluid emission is a challenge, which additionally aims at reducing the drag-out and loss of cutting fluid (Madanchi et al. 2015a).

2.3.3.1 Formation and Characteristics of Emissions

The generated emissions consist of aerosols and vapors. "The term aerosol refers to suspension of liquid or solid particles in a gaseous medium" (Kulkarni et al. 2011). In this context, the aerosol consists of small cutting fluid particles in the air, although it has been investigated that the complex mixture of chemicals in the aerosols may not be exactly identical to the cutting fluid (Byers 2006). In contrast to aerosols that are generated by mechanical stress, vapors are gaseous media that are generated by heat and can also condense to become droplets (DGUV 2012; Kulkarni et al. 2011).

In the literature, different investigations have been performed to identify the mechanisms that lead to the formation of these emissions in machining processes. In this regard, Bell et al. (1999) have identified three main sources: splash, spinoff and evaporation (see Fig. 2.13). Based on conducted turning experiments, splash is described as the effect of momentum transfer in which the fluid hits the workpiece and tool surface

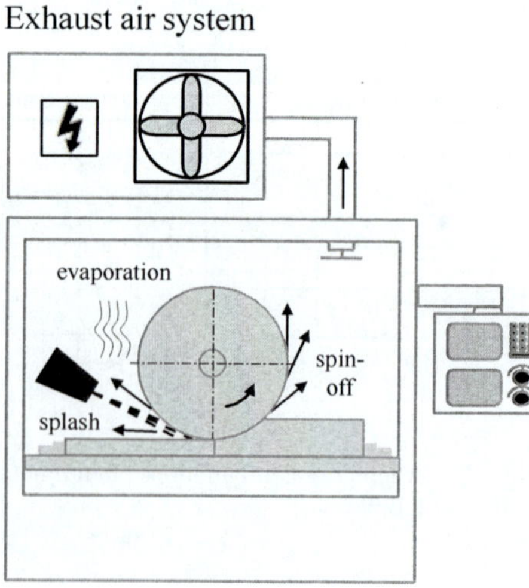

Fig. 2.13 Components of the exhaust air system and mechanisms of emission formation

and then bounces off as droplets. Spin-off is the effect of centrifugal forces generated by the rotation of the workpiece and causes droplets to be released from the fluid-covered surface. These are two mechanical mechanisms that lead to the formation of aerosols, while the third mechanism, evaporation is caused by high temperatures in the contact zone and leads to the formation of vapor. The study by Thornburg and Leith (2000) identified three similar mechanisms, but named them impaction, centrifugal force, and evaporation/condensation. It identified that the largest particles are formed as a result of the centrifugal force and the smallest particles are formed as a result of the evaporation/condensation. Moreover, the formation of cutting fluid emissions also strongly depends on the selected process parameters. Regarding a turning process, a two-part study was conducted focusing on the emission formation mechanisms evaporation and atomization. Yue et al. (2004) first modeled the mass concentration and particle size distribution considering different physical mechanisms and empirical correction factors. The derived models were applied and validated in the second part of the study (Sun et al. 2004). It revealed that the cutting speed is the dominant parameter with the highest impact on the cutting fluid emissions.

However, all these studies focused on the formation of emissions using a conventional turning lathe, which neglected the impact of an exhaust air system of an enclosed machine tool. An approach by Madanchi et al. (2015a) measured the cutting fluid emissions in the exhaust air of a grinding process. The impact of cutting fluid properties such as surface tension and kinematic viscosity as well as process parameters such as cutting speed, volume flow of the cutting fluid and the exhaust air on the formation of emissions were investigated. A measurement in the exhaust air was also conducted by Madanchi et al. (2019a), but focusing on mass concentration and particle size distribution, which demonstrated the highest share of particles at a size of 0.02 μm. Both studies also confirmed the significant impact of the cutting speed on the formation of emissions.

The size of the particles is particularly critical in terms of health, as they deposit in different regions of the respiratory system (extrathoracic, bronchial and alveolar region) depending on their size (see Fig. 2.14). Particles smaller than 0.1 μm are deposited due to diffusion, which describes an irregular zigzag movement of the particles and is also named as Brownian diffusion. The deposition of these particles occurs mainly deeply in the lungs in the alveolar region. Particles with a size between 0.1 and 1 μm are also deposited due to sedimentation and the effect of gravity. Particles within this size range have the lowest deposition rate with partly below 25%. Larger particles above 1 μm are not only deposited due to sedimentation, but also due to inertial impaction when they collide with other objects. In this case, the deposition rate again increases significantly to over 90% (Heyder et al. 1986; Heyder 2004).

The exposure to cutting fluid emissions and their inhalation can therefore be critical and lead to health problems. According to a study by Jaakkola et al. (2009), machine operators have an increased risk of upper-respiratory symptoms, cough, breathlessness and current asthma when they are regularly exposed to cutting fluid emissions. Moreover, emissions can also contain bacteria and endotoxin, which further increase the health risk (Simpson et al. 2003).

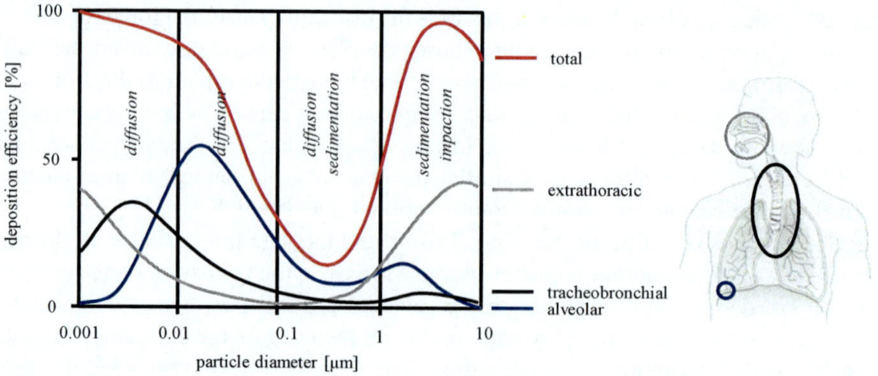

Fig. 2.14 Deposition of particles in the human respiratory system [Madanchi et al. (2019a) based on Heyder et al. (1986), ICRP (1994)]

Against this background, there are emission limits that are enforced by national laws in order to protect occupational health and safety. In Germany, for example, the occupational exposure limit (OEL) for cutting fluids was at 10 mg/m^3 until 2005 (TRGS 900 2000), whereas the German OELs nowadays strongly depends on the actual composition of the emissions (TRGS 900 2006). For simplification the former OEL is still often used in practice (Bannert and Michels 2007). In the USA, the OEL is regulated by the Occupational Safety and Health Administration to a maximum of 5 mg/m^3 (United States Occupational Safety and Health Administration 1989). However, due to the serious health risk, there are also studies in literature suggesting to regulate the OEL for cutting fluids below 0.5 mg/m^3 (Mirer 2010). In order to meet these OELs, it is required to enclose the machine tools and operate them with an effective and suitable exhaust air system.

2.3.3.2 Function and Design of the Exhaust Air System

The control of emissions poses a major challenge for the machining industry when dealing with cutting fluids (Byers 2006). The design of the exhaust air system and the required volume flow is particularly critical as emissions have to be sufficiently removed, whereas the drag-out should be reduced. The flow velocity at the extraction aperture should therefore be less than 4 m/s in order to avoid the removal of larger particles (>300 μm) (VDI 3802-2 2012). Within the extraction duct, the flow velocity should be 12 to 16 m/s and even up to 22 m/s in case of MQL (VDI 3802-2 2012). The determination of the required extraction volume, however, highly depends on the size of machine tool, cutting fluid strategy as well as machining process and parameters. For example, in case of dry machining, high amounts of dust can be generated depending on the workpiece material. Thus, the required extraction volume might be even higher compared to the use of cutting fluid (VDI 3802-2 2012). The selection and dimensioning of the fan is therefore important. In practice, exhaust air systems often operate at a fixed state with constant extraction volume and power demand (Winter 2016). Similar to the pumps, it is also recommended to use frequency control to regulate the fan performance curve depending on the required exhaust air volume (VDMA 2016).

The exhausted air from the machine tool then needs to be filtered and cleaned from the particles by a filtration unit. Generally, four different technologies are available for this purpose, which according to a study had the following shares in Germany in 1991: (I) electrostatic precipitators with about 50%, (II) filtering precipitators with about 37%, (III) centrifugal precipitators with about 6% and (IV) wet precipitators with about 1.5% (BGIA 2006). Electrostatic precipitators are based on the principle that charged particles are deflected within an electrostatic field and thus removed from the air. Despite the extensive use of this technology, it is not recommended for water-based cutting fluids as water in the air can reduce the efficiency of the filter and cause problems by short circuits. In case of filtering precipitators, the exhaust air passes through different filtration media such as filter fleece, demisters or filter mats to mechanically separate the particles. The other two technologies are used comparatively rarely. However, especially in case of explosive and highly flammable emissions, wet precipitators might be recommended, as it is also possible to separate vapors with this technology (VDMA 2016).

2.3.4 Workpiece and Scrap Treatment System

After the actual machining process, chips and workpieces are contaminated and covered with cutting fluids. By using additional peripheral subsystems, they can be processed and, in some cases, cutting fluid can be recovered and returned to the machining system. While a de-oiling process is generally not required in case of MQL or dry machining, the cleaning of the workpiece is also important for these strategies.

2.3.4.1 Workpiece Cleaning

The surface of a workpiece has to be cleaned from any contamination such as chips and cutting fluids that negatively affect the product quality and lead to production failures. This is particularly relevant prior to heat treatment or surface coating, during maintenance as well as for intermediate cleaning between different machining processes (Mang and Dresel 2007).

According to DIN 8592, the cleaning can be divided into six subgroups: blast cleaning, mechanical cleaning, jet cleaning, solvent cleaning, chemical cleaning and thermal cleaning (DIN 8592 2003). However, in order to remove residual chips and cutting fluid, an alkaline immersion or spray cleaning with surfactants is recommended and most commonly used for the cleaning of workpieces in large series

(Kloke 2003). As the cleaning fluid is sprayed with high pressures on the workpiece surface during spray cleaning, immersion cleaning is preferable from an energy perspective. This cleaning technique is sufficient for machined workpieces for further processing, assembly or storage (Amanpour 1995). It can be performed as a single-stage cleaning process or as a multiple-stage cleaning consisting of different immersion and possibly rinsing steps. Often this cleaning technique is also supported by the use of ultrasound (Wullstein 2004). A significant energy demand is connected to the heating of the cleaning fluid to around 15 °C below the flash point of the fluid and to 40–70 °C for the rinsing water (Amanpour 1995). The cleaning fluid is used in a circulating system, but the washed off cutting fluid is lost and must be refilled.

If the same cutting fluid or a multifunctional fluid is used for the following processes, intermediate cleaning may also not even be necessary. In case of MQL or dry machining, the use of cleaning fluids is less common for intermediate cleaning between different machining processes as there is no residual cutting fluid on the workpiece surface. In practice, residual chips and swarf are also removed by surface air cleaners using compressed air. However, as the generation of compressed air is highly energy intensive, this cleaning method is not very efficient (Madanchi et al. 2015b).

2.3.4.2 De-oiling

Chips and swarf generated by the cutting process and separated by the cleaning unit (see Fig. 2.10) represent a valuable resource, which can be recycled and further processed. For this purpose, the quality of the chips is important, which requires mono-material and a low content of oil and of other contaminants such as water or cutting fluid ingredients (Astakhov and Joksch 2012). The de-oiling of chips and swarf is therefore critical when using cutting fluids, while it is not required for MQL or dry machining. In practice, the cutting fluid content of swarf from grinding processes can vary between 10 and 80 mass-%. This large variation is related to the used type of cutting fluid and the size of the chips. Usually, a cutting fluid content of more than 35 mass-% has to be expected (Dettmer 2006). Other literature assumes a cutting fluid content of more than 30 mass-% for low viscosity fluids (Astakhov and Joksch 2012). In case of chips from machining processes with geometrically defined cutting edges, the cutting fluid adhesion is generally lower. Thus, a cutting fluid content of around 20 to 25 mass-% is assumed (Knobloch 1979). For these chips, the use of a chip breaker might be required in order to reduce chips to a suitable size before the de-oiling and recycling process (Astakhov and Joksch 2012).

The actual de-oiling of chips and swarf can be performed by mechanical, thermal or chemical methods. The achievable residual cutting fluid contents with the different methods are summarized in Table 2.5. In practice, the use of mechanical methods and especially centrifugation is most common. It uses the centrifugal force during rotation to separate cutting fluid and chips. Here, recovered cutting fluid can be reused without restrictions and returned to the cutting fluid system. Compared to other methods, it is

Table 2.5 Overview of de-oiling methods [according to Dettmer (2006) based on Hoffmeister and Bock (1997)]

De-oiling method	Cutting fluid content (%)	Cutting fluid reuse	Application details
Centrifugation	15–20	No restriction	Grinding
	1–5	No restriction	Milling, turning
Pressing	15–20	No restriction	Grinding
Washing	<1	Hazardous waste	Grinding
Vacuum thermal	<1	Reducing of additives	Grinding
Extraction with solvent	<1	No restriction	Grinding
Extraction with supercritical CO_2	<1	No restriction	Grinding

less complex with a comparatively low energy demand. Regarding the use of thermal or chemical methods, a high degree of de-oiling can be achieved, but the recovered cutting fluid either cannot be reused or the methods have not been established in practice due to their complexity (Astakhov and Joksch 2012; Dettmer 2006).

2.4 Peripheral Systems Within Machining Process Chains

The design of the peripheral systems for a process chain also depends on the selected cutting fluid strategy. A main aspect in this context is the operation of centralized and decentralized peripheral systems. While a decentralized system serves a single machine tool, the centralized system serves a number of machines. This applies to the cutting fluid as well as the exhaust air systems, as presented in Fig. 2.15.

The decision for a centralized or decentralized system always depends on the specific application and situation including the layout of the manufacturing system, as both concepts have different advantages and disadvantages. A decentralized system offers a high degree of flexibility regarding factory layout and the use of different cutting fluids. In the case of centralized systems, the same cutting fluid has to be used and also the same group of workpiece material has to be machined, as otherwise it can lead to filtration and recycling problems (Joksch 2012). With regard to the exhaust air system, removed emissions can only be returned if the same cutting fluid is used. However, the advantages of the centralized cutting fluid systems are reduced maintenance costs, reduced cutting fluid demand and higher economic efficiency (VDI 3397-2 2007). In this case, only one system needs to be maintained, which requires less personnel. The overall space requirement is also lower (Joksch 2012), but a longer and more complex piping system also leads to higher pressure losses and thus increased energy consumption of the pumps. Centralized cutting fluid systems are often installed in the basement and therefore have to overcome a greater pump

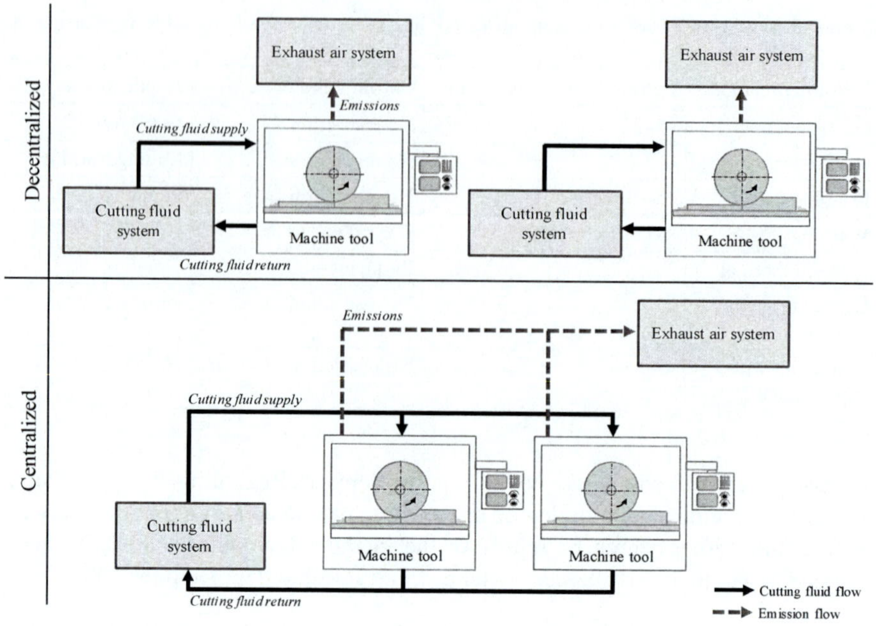

Fig. 2.15 Centralized and decentralized concepts

head and distance to the machine tools. Overall, this again emphasizes that the choice between a centralized or decentralized system is usually case-specific and depends on various aspects.

Besides the cutting fluid and exhaust air system, cleaning units can also be executed as a decentralized system next to the machine tool or as a centralized system. In this case, the utilization and dimensioning of the system is also one of the main factors influencing an environmentally efficient cleaning process (Wilbert et al. 1998). For SMEs with a small scale production, a decentralized system could therefore be more suitable, while centralized systems could be more suitable for large companies with automated process chains and high quantities.

The requirement for cleaning processes within the process chain also depends on the applied cutting fluid strategy. When changing between different machine tools that use different cutting fluids, a cleaning process is necessary to avoid contamination. However, by using a centralized cutting fluid system or the same fluid within a process chain, intermediate cleaning processes can be eliminated (Madanchi et al. 2015b). Further, the application of compatible MFOs enables the reuse of compatible cleaning fluids that are contaminated with compatible cutting fluids and do not have to be disposed (Astakhov and Joksch 2012). It is further possible to eliminate some cleaning processes between different machining processes (Madanchi et al. 2015b) or to reuse a compatible oil adhering to chips and workpieces for the concentration of a compatible emulsion through the use of MFOs (Joksch 2008). This leads to a reduced energy and resource consumption but also shorter cycle times, which probably has an even greater impact.

2.5 Modeling of Machining Systems

In order to systematically analyze the interactions between the presented subsystems as well as to holistically assess the impact of alternative designs and strategies, it is required to transfer them into models and to define performance indicators. Further, improvement approaches are basically presented to increase the efficiency.

2.5.1 Modeling Approach

A model is defined as a "simplified reproduction of a planned or existing system with its processes in a different conceptual or physical system" (VDI 3633 2018). Thus, a system has to be converted into an experimentable model, which can be used to conduct a simulation of experiments in order to gain findings that are transferable to reality. The system should be simplified and limited to the essentials that are required for the specific investigation target. It can be recommended to develop different models and sub-models according to a hierarchical structure or to model specific aspects of the system with a different detail level (VDI 3633-1 2014; VDI 3633 2018). A general modeling approach is presented in Fig. 2.16.

The first step requires the definition of system boundaries and structure. This includes the relation between different system elements and the environment by the definition of input and output values.

The following step is the transformation of the defined system into a model by using a physical or empirical modeling approach. The physical approach uses mathematical functions based on physical/natural laws, while the empirical approach derives a model on the basis of experimental data (Schmitt and Andres 2019). Overall, the physical approach requires a high effort with limited availability of models, but has the main advantage that it is also independent from experimental data and specific application cases. The empirical approach, on the other hand, is characterized by a high dependence on experimental data and limitation of the derived model to the specific process or system. It is therefore hardly possible to apply an empirical model to a different case study. However, its main advantage is the reduced modeling effort and precise description of a specific system (Winter 2016). In practice, the use of semi-empirical models, which are based on physical laws as well as experimental data is therefore also common, as it has the ability to combine the advantages of both approaches (Schmitt and Andres 2019).

In the next step, the derived model is used to perform simulation experiments. By varying the input parameters, the impact on the theoretical results are analyzed. However, the accuracy of these results strongly depends on the model, as it always

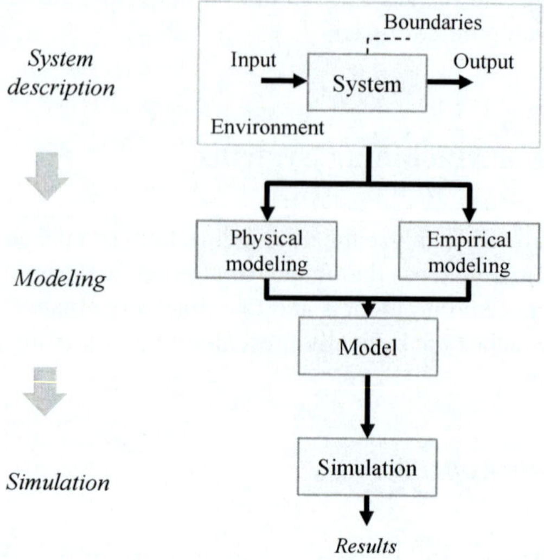

Fig. 2.16 Modeling based on physical and empirical approaches [based on Eisele (2014) Schmitt and Andres (2019)]

represents a simplification and is associated with uncertainties. It is therefore important to validate the quality of the model. In case the agreement of the theoretical results with reality is not sufficient, the model has to be adapted iteratively (VDI 3633-1 2014). This validation can be performed by using different statistical methods. For this purpose, Winter (2016) has, for example, used the coefficient of determination (R^2) to indicate the accuracy and the relative absolute error (E) to indicate the statistical deviation between theoretical and experimental results. In this context, Zirn and Weikert (2006) distinguish for the modeling of manufacturing systems between three accuracy classes:

- Very high level of accuracy with a deviation <5% - achievable with great effort for complex models and useful, if compliance with specific standards or regulations has to be validated
- High level of accuracy with a deviation <30% - achievable with simplified models and sufficient, if few relevant influencing factors are analyzed
- Qualitative level of accuracy with a deviation >30% - achievable with little effort and sufficient, if qualitative conclusions are required.

Eisele (2014) assumes an average accuracy of ±15% of calculated to measured results to be sufficient for his work that focuses on the energy demand of machine tool components.

2.5.2 Modeling of Environmental, Economic and Functional Performance Indicators

In view of the discussed challenges, machining systems have to be designed not only with regard to technological functions but also from an environmental and economic perspective. This places different requirements on the system, which are often difficult to meet simultaneously. Specific correlations are already considered during process design. As an example, it is a general approach to consider the interaction between tool wear, processing time and production costs in order to select a suitable cutting speed (Paucksch et al. 2008). This consideration still needs to be more comprehensive and expanded by assessing the impact on all three perspectives. Several methods and indicators can be used for this purpose as demonstrated in Fig. 2.17. The selection of relevant process parameters in combination with the cutting fluid strategy should therefore be made with regard to the effect on them.

Environmental perspective

By consuming energy and materials as well as generating waste, machining systems have an impact on the environment. This impact can be analyzed by conducting a life cycle assessment (LCA), which is defined as the "compilation and evaluation of the inputs, outputs and the potential environmental impacts of a product system throughout its life cycle" (DIN EN ISO 14040 2009). According to this standard, conducting a LCA comprises four phases. At the phase of goal and scope definition, the intended application and reason for the assessment have to be determined as well as the system boundaries. Furthermore, the functional unit is identified, which is a reference for the performance of a product system. In case of production systems, a specific amount of produced product is often used as a functional unit. The inventory analysis phase focuses on the quantification of all relevant input and output flows along the entire life cycle. The energy and material flows are associated with specific impact categories and converted into equivalents such as CO_2-eq. for global warming potential (GWP) within the impact assessment phase. The interpretation phase then evaluates the findings regarding the defined goal and scope. Overall, the procedure is an iterative process, which does not necessarily require to follow the individual steps in a strict order (Dettmer 2006).

This method is widely used to comparatively assess the environmental impact of different cutting fluid strategies for machining systems with various examples. In view of alternative fluid formulations, Dettmer (2006) evaluated different non-water-miscible cutting fluids based upon vegetable oil and animal fat compared to mineral oil-based fluids in grinding and Winter et al. (2012b) focused on the comparison

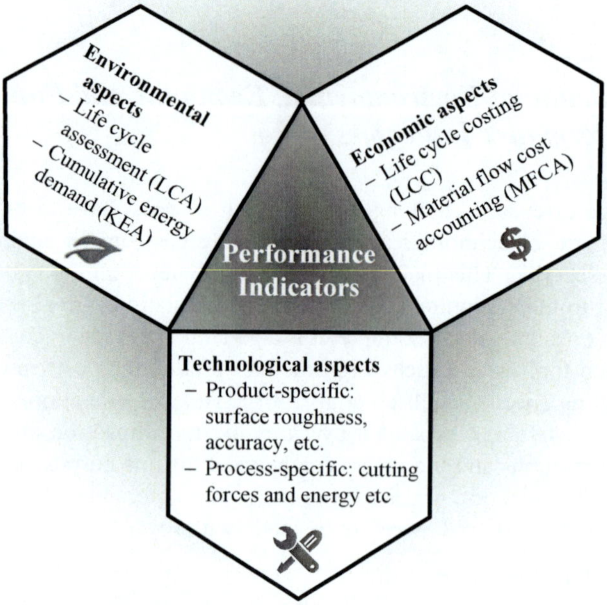

Fig. 2.17 Exemplary methods and performance indicators from a technological, economic and environmental perspective [based on Winter et al. (2013)]

between jatropha oil-based and mineral oil-based emulsions in grinding. With a focus on using alternative amounts of cutting fluids, Fratila (2010) conducted a LCA to compare the impact of flooding and MQL in gear milling. These studies considered the processing of a certain number of workpieces or amount of removed material as the functional unit to compare the strategies. In addition to alternative amounts, Clarens et al. (2008) and Pusavec et al. (2010) also considered alternative media types by assessing the impact of supercritical CO_2 respectively cryogenic cooling. Each study examined a variety of different impact categories, but the consideration of GWP is common to all of them.

Furthermore, the cumulated energy demand (KEA) was also used to evaluate the environmental impact alternative cutting fluid strategies. It demonstrates a representative indicator that focuses on the partial aspect of energy consumption associated with a product system (VDI 4600 2012). The implementation of this method is similar to the LCA and can be considered as "a useful, additional, energy-related 'screening-indicator' to LCA" (Röhrlich et al. 2000). Therefore, the KEA is often used, when a complete LCA may not be performed and energy demand dominates (Dettmer 2006). For example, Herrmann et al. (2017) used the KEA to compare the environmental impact of rapeseed oil-based to mineral oil-based emulsion in milling and Dettmer (2006) also included this approach to the evaluation of alternative fluid formulations in grinding.

Economic perspective

The consumption of materials and energy does not only cause environmental effects, but also economic ones. The method of Life Cycle Costing (LCC) comprises "the "total" costs generated by a system during its service life from the operator point of view" and "aims at optimizing the total costs and yields of a system and of the related activities and processes arising over its life cycle" (VDI 2884 2005). Based on the method, purchase decisions between alternative products can be supported on the one hand, and on the other hand the cost responsibility can be examined over the entire life cycle from the manufacturer's point of view. The use of LCC is advisable when high recurring and follow-up costs are generated, a long service life is estimated or a large quantity has to be purchased (Dettmer 2006; VDI 2884 2005).

In connection with LCA, cost accounting is also performed using an adapted approach based on a material flow analysis. This approach allows both methods to use common models and to create a consistent basis for decision-making (Möller 2000). Combined with the LCA, Dettmer has for example used this material flow approach and conducted a simplified LCC to estimate the production costs and cutting fluid-related costs for the user (Dettmer 2006).

An approach that supports inside into the economic perspective of a system, is the material flow cost accounting (MFCA). It is defined as a method for "quantifying the flows and stocks of materials in processes or production lines in both physical and monetary units" (DIN EN ISO 14051 2011). By the physical and monetary quantification of inputs and outputs of processes or systems, the method aims at a higher transparency of energy and material flows from a cost and environmental perspective. This should also support decision-making in areas such as production planning or process engineering and improve the coordination of the flows (DIN EN ISO 14051 2011). This method is thus designed for the requirements of production processes and focuses on its actual use phase. For this reason, it was used by Winter (2016) to analyze the costs of a grinding system or as a simplified approach by Madanchi et al. (2019c) to compare the machining costs of using alternative cutting fluid strategies.

Technological perspective

For the environmental and economic evaluation of a machining system, data are usually needed which are experimentally gathered. A technological assessment is therefore often integrated and required. However, the performance of experiments is also associated with great effort, which is particularly applicable when empirical models are used instead of physical models. In the context of machining systems, the functional perspective can be either process-specific focusing on the function of the machining process or product-specific focusing on the function of the output respectively the workpiece.

For the evaluation of the machining process, temperature, vibration, cutting force or cutting energy are examples for process-specific indicators, which give an indication of the process quality. Particularly the use of the cutting force is common. The indicators can be determined by using different approaches such as numerical methods e.g. finite element method (FEM) or empirical methods. However, in terms of cutting forces, the use of FEM is rather complex compared to the Kienzle model which is a common empirical approach based on the specific force to remove a specific chip size (Kienzle 1952). For the calculation, empirical values are available in the literature for a large number of different tool and material combinations, which simplifies the application compared to other methods. It can be applied to turning, milling or grinding process and also used to determine the cutting energy. This approach was, for example, used by Eisele (2014), Kuhrke (2011), Schrems (2014).

Regarding product-specific indicators, the focus is on the quality of the produced workpiece. This can be expressed by indicators such as the workpiece accuracy, metallurgical damage or surface roughness. Further, the surface integrity combines external aspects of the workpiece surface (e.g. roughness) as well as internal aspects (e.g. residual stress). However, as there is a variety of different influencing parameters in machining, it is extremely difficult to predict these properties and it already required great effort to determine a relationship between process parameters and surface roughness (Davim 2010). Due to the extensive investigations, a variety of different models for the determination of surface roughness have been developed in the literature for different machining processes (e.g. Khare and Agarwal 2015; Nalbant et al. 2007; Ståhl et al. 2011; Su et al. 2019; Wang et al. 2019).

2.6 Preliminary Conclusion

This chapter demonstrates the interactions within machining systems consisting of machining processes and peripheral subsystems. It further considers the complex challenge of considering the impact of cutting fluid strategies on it. The complexity results from the versatile influence on various subsystems and components from a technological, environmental and economic perspective. This poses the risk of problem shifting and therefore requires an integrated approach to support decision-making. The approach should follow the efficiency strategy in order to reduce the overall energy and resource consumption.

In view of the existing knowledge and requirements, this includes technological and procedural as well as methodological challenges:

- With respect to the current challenges, machining systems have to be designed not only with regard to technical functions, but also from an environmental and economic perspective. This places different requirements on the system, which are often difficult to meet simultaneously.

- Certain interactions between individual subsystems (e.g. machining process and tool preparation) are already considered, for example, to determine process parameters from an economic perspective. These interactions are, however, limited and need to be extended with regard to cutting fluids.
- In order to support an integrated planning relevant interactions and flows between the subsystems as well as their parameters and properties need to be specified.
- The reproduction of a system in a model can be achieved by an empirical or physical approach. This requires simplifications in order to reduce the effort but is also associated with uncertainties. A combination of both approaches are semi-empirical models.
- With a large number of different influencing parameters on the one hand and strong simplifications on the other, it is necessary that the models achieve a certain accuracy. In order to ensure that theoretical results are transferable to reality it is required to validate the quality of the model.
- Certain components of the subsystems (e.g. pumps) are particularly critical. It is thus required to model these components at a higher level of detail than others. Further, models and sub-models should be developed to a hierarchical structure.

The presented challenges need to be considered for the planning and operation of machining systems. In order to support an integrated decision making it is therefore important to address the presented challenges.

References

Amanpour A (1995) Reinigung mit Kohlenwasserstoffen und Wasser, vol 469. Expert-Verlag, Renningen-Malmsheim, Germany

Astakhov VP, Joksch S (eds) (2012) Metalworking fluids for cutting and grinding: fundamentals and recent advances. Woodhead Publishing in materials. Woodhead Publishing, Cambridge, Philadephia

Bannert P, Michels P (2007) Lufttechnische Maßnahmen bei Tätigkeiten mit Kühlschmierstoffen. Berufsgenossenschaft der Feinmechanik und Elektrotechnik, Köln, Germany

Bell DD, Chou J, Nowag L, Liang SY (1999) Modeling of the environmental effect of cutting fluid. Tribol Trans 42:168–173. https://doi.org/10.1080/10402009908982204

BGIA (ed) (2006) Absaugen und Abscheiden von Kühlschmierstoffemissionen. Hauptverband der gewerblichen Berufsgenossenschaften (HVBG), Sankt Augustin, Germany

Boothroyd G, Knight WA (2006) Fundamentals of machining and machine tools, 3rd edn. Mechanical engineering, vol 69. CRC Taylor & Francis, Boca Raton

Brecher C, Weck M (2019) Werkzeugmaschinen Fertigungssysteme 1, 9th edn. VDI-Buch. Springer Vieweg, Berlin, Germany

Brecher C, Triebs J, Heyers C, Jaspers D (2012) Effizienzsteigerung von Werkzeugmaschinen durch Optimierung der Technologien zum Komponentenbetrieb - EWOTeK. Apprimus Verlag, Aachen, Germany

Brinksmeier E, Bohling A (2008) Beherrschung des Kühlschmierstoffeinsatzes beim Schleifen. Moderne Schleiftechnologie Und Feinstbearbeitung 7:1–26

Brinksmeier E, Heinzel C, Wittmann M (1999) Friction, cooling and lubrication in grinding. CIRP Ann 48:581–598

Brinksmeier E, Meyer D, Huesmann-Cordes AG, Herrmann C (2015) Metalworking fluids—mechanisms and performance. CIRP Ann 64:605–628. https://doi.org/10.1016/j.cirp.2015.05.003
Byers JP (2006) Metalworking fluids, 2nd edn. CRC Press, Hoboken
Clarens AF, Zimmerman JB, Keoleian GA, Hayes KF, Skerlos SJ (2008) Comparison of life cycle emissions and energy consumption for environmentally adapted metalworking fluid systems. Environ Sci Technol 42:8534–8540. https://doi.org/10.1021/es800791z
Davim JP (2010) Surface integrity in machining. Springer, London
Denkena B, Tönshoff HK (2011) Spanen. Springer, Berlin, Heidelberg
Dettmer T (2006) Nichtwassermischbare Kühlschmierstoffe auf Basis nachwachsender Rohstoffe. Dr.-Ing. dissertation, Technische Universität Braunschweig. Vulkan Verlag, Essen, Germany
DGUV (2012) Brand- und Explosionsschutz an Werkzeugmaschinen. BGI/GUV-I 719. Deutsche Gesetzliche Unfallversicherung e.V., Berlin, Germany
Dietrich J (2016) Praxis der Zerspantechnik. Springer, Wiesbaden, Germany
DIN 51385 (2013) Schmierstoffe - Bearbeitungsmedien für die Umformung und Zerspanung von Werkstoffen - Begriffe. DIN, Beuth Verlag GmbH, Berlin, Germany
DIN 8580 (2020) Manufacturing processes—terms and definitions, division. DIN, Beuth Verlag GmbH, Berlin, Germany
DIN 8589-0 (2003) Manufacturing processes chip removal. DIN, Beuth Verlag GmbH, Berlin, Germany
DIN 8592 (2003) Fertigungsverfahren Reinigen. DIN, Beuth Verlag GmbH, Berlin, Germany
DIN EN 60034-30-1 (2014) Drehende elektrische Maschinen - Wirkungsgrad-Klassifizierung von netzgespeisten Drehstrommotoren. DIN, Beuth Verlag GmbH, Berlin, Germany
DIN EN ISO 14040 (2009) Environmental management—life cycle assessment—principles and framework. DIN, Beuth Verlag GmbH, Berlin, Germany
DIN EN ISO 14051 (2011) Environmental management—material flow cost accounting—general framework. DIN, Beuth Verlag GmbH, Berlin, Germany
DIN EN ISO 9000 (2015) Quality management systems—fundamentals and vocabulary. DIN, Beuth Verlag GmbH, Berlin, Germany
Döbbeler B, Klocke F, Lung D (2015) Methodology of process oriented dimensioning of cooling lubricant pressure and volume flow for increasing energy efficiency. In: 22th CIRP international conference on life cycle engineering, Sydney, Australia, vol 29, pp 347–353
DOE (2014) Improving motor and drive system performance: a sourcebook for industry. U.S. Department of Energy, Advanced Manufacturing Office, Washington DC, USA
Dröder K, Karpuschewski B, Uhlmann E, Arrabiyeh PA, Berger D, Busemann S, Hartig J, Madanchi N, Mahlfeld G, Sommerfeld C (2020) A comparative analysis of ceramic and cemented carbide end mills. Prod Eng Res Devel 14:355–364. https://doi.org/10.1007/s11740-020-00966-9
Eisele C (2014) Simulationsgestützte Optimierung des elektrischen Energiebedarfs spanender Werkzeugmaschinen. Dr.- Ing. dissertation, TU Darmstadt. Shaker Verlag, Aachen, Germany
Engineer F, Guo C, Malkin S (1992) Experimental measurement of fluid flow through the grinding zone. J Eng Ind 114:61–66. https://doi.org/10.1115/1.2899759
Fratila D (2010) Macro-level environmental comparison of near-dry machining and flood machining. J Clean Prod 18:1031–1039. https://doi.org/10.1016/j.jclepro.2010.01.017
Goindi GS, Sarkar P (2017) Dry machining: a step towards sustainable machining—challenges and future directions. J Clean Prod 165:1557–1571. https://doi.org/10.1016/j.jclepro.2017.07.235
Graham W, Whiston MG (1978) Some observations of through-wheel coolant application in grinding. Int J Mach Tool Des Res 18:9–18
Gviniashvili V (2003) Fluid application system optimisation for high speed grinding. Ph.D. thesis, Liverpool John Moores University, UK
Heinzel C (1999) Methoden zur Untersuchung und Optimierung der Kühlschmierung beim Schleifen. Dr.- Ing. dissertation, Universität Bremen. Shaker Verlag, Aachen, Germany
Heinzel C (2016) Energieeffizienter KSS-Einsatz beim Schleifen. Schlussbericht IGF-Vorhaben 17863, Universität Bremen, Bremen, Germany

Helpertz M, Wenderott D, Wagner P (2010) Energieeffizienter Betrieb von Schraubenspindelpumpen an Werkzeugmaschinen. 8. VDI-Fachtagung Schraubenmaschinen, Dortmund, Germany VDI-Verlag, pp 137–149

Herrmann C, Thiede S, Zein A, Ihlenfeldt S, Blau P (2009) Energy efficiency of machine tools: extending the perspective. In: 42th CIRP conference on manufacturing systems, Grenoble, France

Herrmann C, Madanchi N, Winter M, Öhlschläger G, Greßmann A, Zettl E, Schwengers K, Lange U (2017) Ökologische und ökonomische Bewertung des Ressourcenaufwands - Wassermischbare Kühlschmierstoffe. Technical Report, VDI Zentrum Ressourceneffizienz GmbH (VDI ZRE)

Hesselbach J (2012) Energie- und klimaeffiziente Produktion. Springer Vieweg, Wiesbaden, Germany

Heyder J (2004) Deposition of inhaled particles in the human respiratory tract and consequences for regional targeting in respiratory drug delivery. Proc Am Thorac Soc 1:315–320. https://doi.org/10.1513/pats.200409-046TA

Heyder J, Gebhart J, Rudolf G, Schiller CF, Stahlhofen W (1986) Deposition of particles in the human respiratory tract in the size range 0.005–15 μm. J Aerosol Sci 17:811–825

Hoffmeister H-W, Bock R (1997) Umweltverträglichere Schleifbearbeitung durch Ressourcenschonung. VDI-Berichte 1339

ICRP (1994) ICRP model: human respiratory tract model for radiological protection. International Commission for Radiological Protection, vol 66. ICRP Publication, pp 1–482

ISO 3685 (1977) Tool-life testing with single-point turning tools. International Organization for Standardization

Jaakkola MS, Suuronen K, Luukkonen R, Järvelä M, Tuomi T, Alanko K, Mäkelä EA, Jolanki R (2009) Respiratory symptoms and conditions related to occupational exposures in machine shops. Scand J Work Environ Health 35:64–73

Joksch S (2008) Abfallfreie Kurbelwellenfertigung – Ressourcenschonung und Kostenoptimierung durch Multifunktionsöle und kompatible Metallbearbeitungsflüssigkeiten. In: Denkena B (ed) Werkzeuge für die Zerspanung: Entwicklung, Prozesskette, Einsatz; Begleitband zum Seminar. PZH Verlag, Garbsen

Joksch S (2012) Circulation systems for metalworking fluids. In: Astakhov VP, Joksch S (eds) Metalworking fluids for cutting and grinding: fundamentals and recent advances. Woodhead Publishing, Cambridge, Philadephia

Khare SK, Agarwal S (2015) Predictive modeling of surface roughness in grinding. In: 15th CIRP conference on modelling of machining operations, Karlsruhe, Germany, vol 31, pp 375–380. https://doi.org/10.1016/j.procir.2015.04.092

Kienzle O (1952) Die Bestimmung von Kräften und Leistungen an spanenden Werkzeugen und Werkzeugmaschinen. VDI-Z, pp 299–305

Kirsch B, Effgen C, Büchel M, Aurich JC (2014) Comparison of the embodied energy of a grinding wheel and an end mill. In: 21th CIRP international conference on life cycle engineering, Trondheim, Norway, vol 15, pp 74–79. https://doi.org/10.1016/j.procir.2014.06.037

Klocke F (2009) Manufacturing processes 2: grinding, honing, lapping. Springer, Berlin, Heidelberg

Klocke F (2011) Manufacturing processes 1: cutting. Springer, Berlin, Heidelberg

Klocke F, Eisenblätter G (1997) Dry cutting. CIRP Ann 46:519–526. https://doi.org/10.1016/S0007-8506(07)60877-4

Kloke U (2003) Auslegung von Bauteilreinigungsanlagen mit Hilfe eines Fachinformationssystems. Dr.- Ing. dissertation, TU Dortmund. Maschinenelemente Verlag, Dortmund, Germany

Knobloch H (1979) Kühlschmierstoffpflege in der Praxis. Lexika Verlag, Grafenau, Württ, Germany

Koch T (2007) Auswirkungen des mikrobiellen Befalls von wassergemischten Kühlschmierstoffen auf das Zerspanergebnis. Research report AiF 14056 N\1, Universität Bremen

KSB (2005) Auslegung von Kreiselpumpen: Technische information, 5th edn. KSB AG, Frankenthal, Germany

Kuhrke B (2011) Methode zur Energie- und Medienbedarfsbewertung spanender Werkzeugmaschinen. Dr.- Ing. dissertation, TU Darmstadt. epubli, Berlin, Germany

Kulkarni P, Baron PA, Willeke K (2011) Aerosol measurement: principles, techniques, and applications. Wiley, Hoboken

LfU (2009) Leitfaden für effiziente Energienutzung in Industrie und Gewerbe. Bayerisches Landesamt für Umwelt, Augsburg, Germany

Madanchi N, Winter M, Herrmann C (2015a) Cutting fluid drag-out and exhaust air in grinding processes: influence on the eco-efficiency. In: 22th CIRP international conference on life cycle engineering, Sydney, Australia, vol 29, pp 329–334. https://doi.org/10.1016/j.procir.2015.02.054

Madanchi N, Kurle D, Winter M, Thiede S, Herrmann C (2015b) Energy efficient process chain: the impact of cutting fluid strategies. In: 22th CIRP international conference on life cycle engineering, Sydney, Australia, vol 29, pp 360–365. https://doi.org/10.1016/j.procir.2015.02.056

Madanchi N, Winter M, Thiede S, Herrmann C (2017) Energy efficient cutting fluid supply: the impact of nozzle design. In: 24th CIRP international conference on life cycle engineering, Kamakura, Japan, vol 61, pp 564–569. https://doi.org/10.1016/j.procir.2016.11.192

Madanchi N, Leiden A, Winter M, Asbach C, Lindermann J, Herrmann C, Thiede S (2019a) Cutting fluid emissions in grinding processes: influence of process parameters on particle size and mass concentration. Int J Adv Manuf Technol 101:773–783. https://doi.org/10.1007/s00170-018-2934-5

Madanchi N, Zellmer S, Winter M, Flach F, Garnweitner G, Herrmann C (2019b) Investigation on the effects of nanoparticles on cutting fluid properties and tribological characteristics. Int J Precis Eng Manuf-Green Technol 6:433–447. https://doi.org/10.1007/s40684-019-00053-0

Madanchi N, Thiede S, Gutowski TG, Herrmann C (2019c) Modeling the impact of cutting fluid strategies on environmentally conscious machining systems. In: 26th CIRP international conference on life cycle engineering, West Lafayette, vol 80, pp 150–155. https://doi.org/10.1016/j.procir.2019.01.068

Malkin S, Guo C (2008) Grinding technology: theory and applications of machining with abrasives, 2nd edn. Industrial Press, New York

Mang T, Dresel W (2007) Lubricants and lubrication. Wiley-VCH Verlag, Weinheim, Germany

Marinescu ID, Rowe WB, Dimitrov B, Ohmori H (2004) Tribology of abrasive machining processes. William Andrew, Norwich

Marinescu ID, Hitchiner MP, Uhlmann E, Rowe WB, Inasaki I (2007) Handbook of machining with grinding wheels. CRC Press Taylor & Francis Group, Boca Raton, USA

Meister M (2011) Vademecum des Schleifens. Hanser, München, Germany

Mirer FE (2010) New evidence on the health hazards and control of metalworking fluids since completion of the OSHA advisory committee report. Am J Ind Med 53:792–801. https://doi.org/10.1002/ajim.20853

Misra KB (ed) (2008) Handbook of performability engineering. Shaker Verlag, London, UK

Möller A (2000) Grundlagen stoffstrombasierter betrieblicher Umweltinformationssysteme. Dr. rer.nat. dissertation, Universität Hamburg. Projekt-Verlag, Bochum, Germany

Möller UJ, Nassar J (2002) Schmierstoffe im Betrieb, 2nd edn. Springer, Berlin, Heidelberg, Germany

Müller B (2004) Thermische Analyse des Zerspanens metallischer Werkstoffe bei hohen Schnittgeschwindigkeiten. Dr.-Ing. dissertation, RWTH Aachen, Aachen, Germany

Nalbant M, Gökkaya H, Sur G (2007) Application of Taguchi method in the optimization of cutting parameters for surface roughness in turning. Mater Des 28:1379–1385. https://doi.org/10.1016/j.matdes.2006.01.008

Paucksch E, Holsten S, Linß M, Tikal F (2008) Zerspantechnik: Prozesse, Werkzeuge, Technologien, 12th edn. Vieweg + Teubner, Wiesbaden, Germany

Priarone PC, Robiglio M, Settineri L, Tebaldo V (2016) Modelling of specific energy requirements in machining as a function of tool and lubricoolant usage. CIRP Ann 65:25–28. https://doi.org/10.1016/j.cirp.2016.04.108

Pusavec F, Krajnik P, Kopac J (2010) Transitioning to sustainable production—part I: application on machining technologies. J Clean Prod 18:174–184. https://doi.org/10.1016/j.jclepro.2009.08.010

Rabenstein A, Koch T, Remesch M, Brinksmeier E, Kuever J (2009) Microbial degradation of water miscible metal working fluids. Int Biodeterior Biodegrad 63:1023–1029. https://doi.org/10.1016/j.ibiod.2009.07.005

Rahäuser R (2015) Energieeffizienter Betrieb von Kühlschmierstoff-Reinigungsanlagen mit Anschwemmfiltern. Dr.- Ing. dissertation, Universität Stuttgart. Fraunhofer Verlag, Stuttgart, Germany

Reik M, Wnuk R (2014) Filtration of metal machining fluids. In: Mang T (ed) Encyclopedia of lubricants and lubrication. Springer, Berlin, Heidelberg

Röhrlich M, Mistry M, Martens PN, Buntenbach S, Ruhrberg M, Dienhart M, Briem S, Quinkertz R, Alkan Z, Kugeler K (2000) A method to calculate the cumulative energy demand (CED) of lignite extraction. Int J Life Cycle Assess 5:369–373. https://doi.org/10.1007/BF02978675

Rowe WB (2009) Principles of modern grinding technology. William Andrew, Oxford, UK

Sales WF, Schoop J, Jawahir IS (2017) Tribological behavior of PCD tools during superfinishing turning of the Ti6Al4V alloy using cryogenic, hybrid and flood as lubri-coolant environments. Tribol Int 114:109–120. https://doi.org/10.1016/j.triboint.2017.03.038

Salonitis K, Ball P (2013) Energy efficient manufacturing from machine tools to manufacturing systems. In: 46th CIRP conference on manufacturing systems, vol 7, pp 634–639. https://doi.org/10.1016/j.procir.2013.06.045

Schmitt TL, Andres M (2019) Methoden zur Modellbildung und Simulation mechatronischer Systeme. Springer, Wiesbaden, Germany

Schrems S (2014) Methode zur modellbasierten Integration des maschinenbezogenen Energiebedarfs in die Produktionsplanung. Dr.- Ing. dissertation, TU Darmstadt. Shaker Verlag, Aachen, Germany

Seidel B, Heinzel C, Meyer D, Geilert P, Karpuschewski B (2020) Sustainable machining by energy- and resource-efficient application of metalworking fluids. In: 17th global conference on sustainable manufacturing, procedia manufacturing, vol 43, pp 151–158

Sigloch H (2008) Technische Fluidmechanik, 6th edn. Springer, Berlin, Heidelberg, Germany

Simpson AT, Stear M, Groves JA, Piney M, Bradley SD, Stagg S, Crook B (2003) Occupational exposure to metalworking fluid mist and sump fluid contaminants. Ann Occup Hyg 47:17–30. https://doi.org/10.1093/annhyg/meg006

Ståhl J-E, Schultheiss F, Hägglund S (2011) Analytical and experimental determination of the Ra surface roughness during turning. In: 1st CIRP conference on surface integrity, procedia engineering, vol 19, pp 349–356. https://doi.org/10.1016/j.proeng.2011.11.124

Su H, Yang C, Gao S, Fu Y, Ding W (2019) A predictive model on surface roughness during internal traverse grinding of small holes. Int J Adv Manuf Technol 103:2069–2077. https://doi.org/10.1007/s00170-019-03643-z

Suda S, Yokota H, Inasaki I, Wakabayashi T (2002) A synthetic ester as an optimal cutting fluid for minimal quantity lubrication machining. CIRP Ann 51:95–98. https://doi.org/10.1016/S0007-8506(07)61474-7

Sun J, Ju C, Yue Y, Gunter KL, Michalek DJ, Sutherland JW (2004) Character and behavior of mist generated by application of cutting fluid to a rotating cylindrical workpiece, part 2: experimental validation. J Manuf Sci Eng 126:426. https://doi.org/10.1115/1.1765151

Suuronen K (2009) Metalworking fluids—allergens, exposure, and skin and respiratory effects. PhD thesis, University of Kuopio, Kuopio, Finland

Thornburg J, Leith D (2000) Size distribution of mist generated during metal machining. Appl Occup Environ Hyg 15:618–628. https://doi.org/10.1080/10473220050075626

TRGS 900 (2000) Arbeitsplatzgrenzwerte in der Luft am Arbeitsplatz „Luftgrenzwerte". Bundesanstalt für Arbeitsschutz und Arbeitsmedizin, Dortmund, Germany

TRGS 900 (2006) Arbeitsplatzgrenzwerte in der Luft am Arbeitsplatz „Luftgrenzwerte". Bundesanstalt für Arbeitsschutz und Arbeitsmedizin, Dortmund, Germany

Triebs JB (2015) Analyse, Optimierung und Simulation des Energieumsatzes dezentraler Fluidkreisläufe von Werkzeugmaschinen. Dr.-Ing. Dissertation, RWTH Aachen. Apprimus Verlag, Aachen, Germany

Triesch J (2010) Reinigung von Kühlschmierstoffen: Konzepte, Methoden und Hinweise für den Praktiker, 1st edn. Expert Verlag, Renningen, Germany

United States Occupational Safety and Health Administration (1989) 1910.1000 Table Z-1: Table Z-1 limit for air contaminants. United States Occupational Safety and Health Administration, United States Department of Labor, USA

VDI 2884 (2005) Purchase, operating and maintenance of production equipment using Life Cycle Costing (LCC). Verein Deutscher Ingenieure, Berlin, Germany

VDI 3035 (2008) Design of machine tools, production lines and peripheral equipment for the use of metalworking fluids. Verein Deutscher Ingenieure, Berlin, Germany

VDI 3397-1 (2007) Metalworking fluids. Verein Deutscher Ingenieure, Berlin, Germany

VDI 3397-2 (2007) Maintenance of metalworking fluids. Verein Deutscher Ingenieure, Berlin, Germany

VDI 3633 (2018) Simulation of systems in materials handling, logistics and production—terms and definitions. Verein Deutscher Ingenieure, Berlin, Germany

VDI 3633-1 (2014) Simulation of systems in materials handling, logistics and production—fundamentals. Verein Deutscher Ingenieure, Berlin, Germany

VDI 3802-2 (2012) Air conditioning systems for factories. Verein Deutscher Ingenieure, Berlin, Germany

VDI 4600 (2012) Cumulative energy demand (KEA). Verein Deutscher Ingenieure, Berlin, Germany

VDMA (2016) Separating and filtering metalworking fluids. Air Pollution Control, Frankfurt am Main, Germany

Wang R, Wang B, Barber G, Gu J, Schall J (2019) Models for prediction of surface roughness in a face milling process using triangular inserts. Lubricants 7:9. https://doi.org/10.3390/lubricants7010009

Webster JA, Cui C, Mindek RB, Lindsay R (1995) Grinding fluid application system design. CIRP Ann 44:333–338. https://doi.org/10.1016/S0007-8506(07)62337-3

Weinert K, Inasaki I, Sutherland JW, Wakabayashi T (2004) Dry machining and minimum quantity lubrication. CIRP Ann 53:511–537. https://doi.org/10.1016/S0007-8506(07)60027-4

Wilbert H-P, Kreisel G, Goldhan G (1998) Ganzheitliche Bilanzierung/Bewertung von Reinigungs-/Vorbehandlungstechnologien in der Oberflächenbehandlung: Forschungsbericht Ganzheitliche Bilanzierung/Fallstudien. Bundesministerium für Bildung und Forschung (BMBF)

Winter M (2016) Eco-efficiency of grinding processes and systems. Dr.-Ing. dissertation, Technische Universität Braunschweig. Springer, Switzerland

Winter M, Bock R, Herrmann C (2012a) Investigation of a new ecologically benign metalworking fluid in abrasive machining processes to substitute mineral oil based fluids. In: 5th CIRP conference on high performance cutting, vol 1, pp 393–398. https://doi.org/10.1016/j.procir.2012.04.070

Winter M, Öhlschläger G, Dettmer T, Ibbotson S, Kara S, Herrmann C (2012b) Using Jatropha oil based metalworking fluids in machining processes: a functional and ecological life cycle evaluation. In: 19th CIRP international conference on life cycle engineering, Berkeley, USA, pp 311–317. https://doi.org/10.1007/978-3-642-29069-5_53

Winter M, Bock R, Herrmann C (2013) Investigation of a new polymer-water based cutting fluid to substitute mineral oil based fluids in grinding processes. CIRP J Manuf Sci Technol 6:254–262. https://doi.org/10.1016/j.cirpj.2013.07.003

Winter M, Ibbotson S, Kara S, Herrmann C (2015) Life cycle assessment of cubic boron nitride grinding wheels. J Clean Prod 107:707–721. https://doi.org/10.1016/j.jclepro.2015.05.088

Wittel H, Jannasch D, Voßiek J, Spura C (2017) Roloff/Matek Maschinenelemente. Springer, Wiesbaden, Germany

Wittmann M (2007) Bedarfsgerechte Kühlschmierung beim Schleifen. Dr.- Ing. dissertation, Universität Bremen. Shaker Verlag, Aachen, Germany

Wullstein M (2004) Auswahl und optimale Auslegung industrieller Bauteilreinigungsanlagen. Dr.-Ing. dissertation, Universität Dortmund. Maschinenelemente Verlag, Dortmund, Germany

Yue Y, Sun J, Gunter KL, Michalek DJ, Sutherland JW (2004) Character and behavior of mist generated by application of cutting fluid to a rotating cylindrical workpiece, part 1: model development. J Manuf Sci Eng 126:417. https://doi.org/10.1115/1.1765150

Zirn O, Weikert S (2006) Modellbildung und Simulation hochdynamischer Fertigungssysteme. Springer, Berlin, Heidelberg, Germany

Zwingmann G (1979) Kühlschmierstoffe für die spanende Metallbearbeitung. Werkstatt Und Betrieb 112:409–414

Chapter 3
State of Research

This chapter focuses on the state of research regarding the configuration of machining systems in view of technological, environmental and economic requirements. As a first step, central perspectives are presented for the limitation of relevant approaches, which are classified in a citation network and provide a general overview (Sect. 3.1). Further, additional criteria are developed to analyze and compare selected approaches within the context of energy and resource efficient machining systems (Sect. 3.2). On the basis of these criteria, a detailed review of the selected literature is conducted to provide a comparative overview. The research gap revealed by the comparative overview demonstrates the need for further research (Sect. 3.3).

3.1 Procedure and Overview of Research Approaches

As there exists a large variety of different research approaches in literature dealing with sustainable manufacturing and machining systems, it is necessary to filter and to identify relevant approaches. For this purpose, the general procedure shown in Fig. 3.1 is followed. In the first step, different databases such as Scopus and Google Scholar are searched for literature using keywords in the field of sustainable machining systems and cutting fluid strategies. This literature screening results in a broad overview, which is further narrowed down in terms of relevant approaches.

For literature limitation, relevant research approaches need to consider a machining system or its relevant subsystems as described in the previous chapter. This comprises on the one hand the machining process itself and on the other hand also the peripheral subsystems such as the cutting fluid system, the exhaust air system and/or the workpiece and scrap treatment system. With regard to a sustainable planning and operation of the machining system, only approaches that consider not only the technological dimension but also the environmental and/or economic dimension are relevant for this work. In terms of applied methods, this also implies that

N. Madanchi, *Model Based Approach for Energy and Resource Efficient Machining Systems*, Sustainable Production, Life Cycle Engineering and Management,
https://doi.org/10.1007/978-3-030-87540-4_3

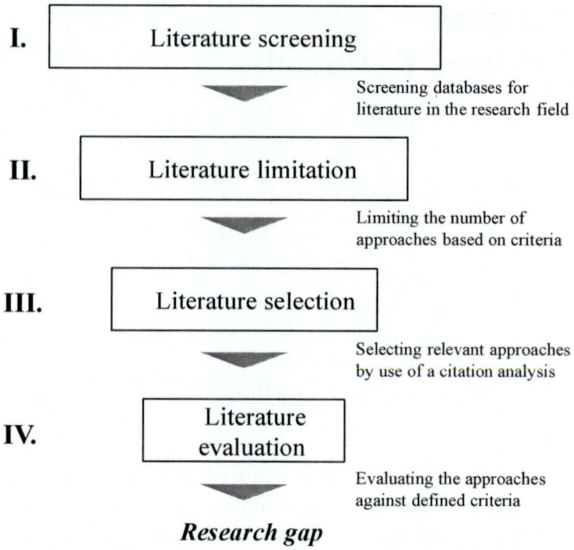

Fig. 3.1 General procedure for the literature review

relevant approaches describe, model, evaluate or optimize the energy and resource flows between the elements of a machining system. This results in three central perspectives, which have to be included within relevant research approaches:

- *Machining system perspective* - research approaches need to focus on the machining system and/or its relevant subsystems.
- *Sustainability perspective* - research approaches need to consider the economical and/or environmental dimension additional to the technological dimension.
- *Methodical perspective* - research approaches need to describe, model, evaluate and/or improve the considered machining system/subsystems.

Thus, research approaches are not taken into account if they only focus on the technological perspective or present a review of existing literature without an independent and innovative approach.

Based on these perspectives, the general literature overview is limited to 70 relevant research approaches. As shown in Fig. 3.2, these approaches are further analyzed within a citation network using the visualization software NodeXL®. For each approach, the size of the symbol indicates the total number of citations and the connections between the approaches represent the information flows from earlier approaches to later ones. This allows an evaluation of the general scientific importance as well as the relative importance of an approach within the network. Approaches that have rarely been cited so far, may still have high relative relevance if they have large information flows within the network. This visualization further categorizes the approaches according to the year of publication, the type of publication and the respective focus level of the approach. In accordance with Sect. 2.3, the *process level* comprises approaches, which mainly focus on the interaction between tool and workpiece as well as the impact of process parameters on the performance.

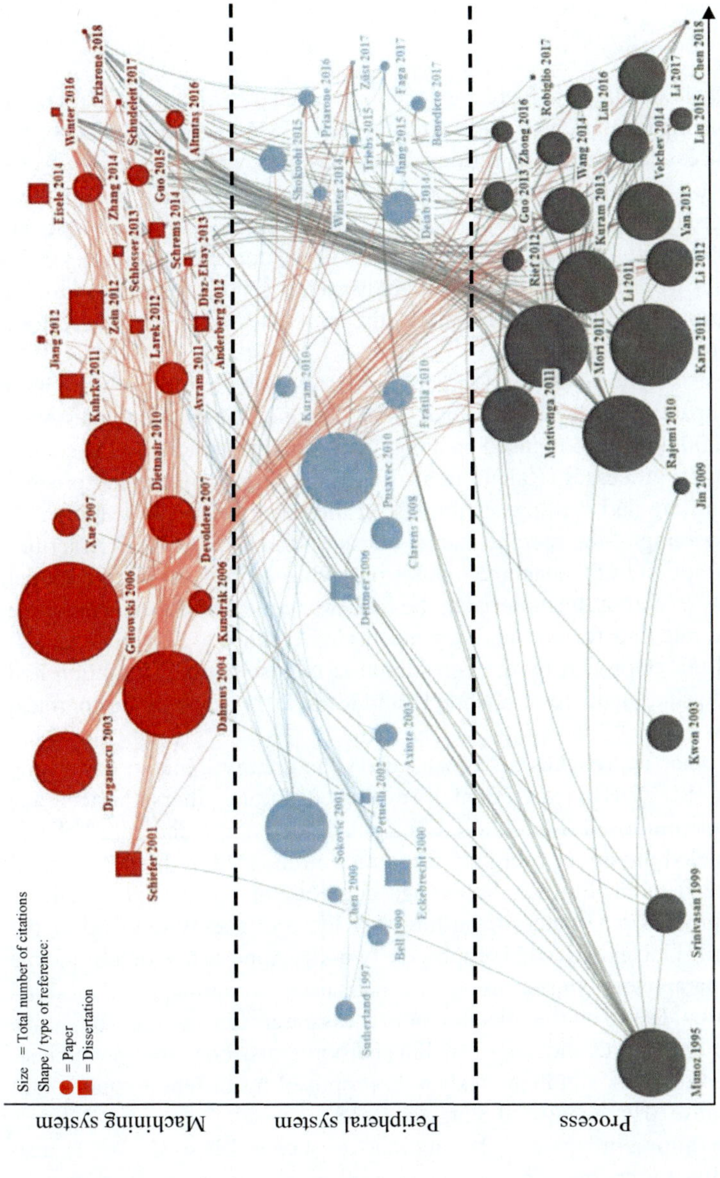

Fig. 3.2 Visualization of the citation network of selected research approaches in the field[1]

[1] Data based on Scopus and Google Scholar last accessed January 2019.

The *peripheral subsystem level* explicitly focuses on its design and performance under consideration of different strategies. The *machining system level* focuses on the process as well as the peripheral subsystems as a combination and part of the entire machining system. The approaches for each focus level are briefly presented in the following.

Process level

At process level, it is striking that there are earlier approaches, which are still relevant and still have an impact within the scientific literature. Munoz and Sheng (1995) modeled the process mechanics of turning and milling as well as relevant waste streams in order to assess the impact of process parameters on the environment. The approach presented by Srinivasan and Sheng (1999) aimed to integrate environmental factors in process planning as well by modeling the energy demand, machining time, waste streams and workpiece quality. Kwon and Fischer (2003) focused on a tool life model to improve the use of tools for machining processes with regard to surface roughness and production cost.

Furthermore, it is noticeable that a large number of research approaches with a high scientific impact were developed since 2010. Three of these approaches were cited more than 200 times and focused on an optimal tool life to increase the energy efficiency of turning processes (Rajemi et al. 2010), the impact of varying process parameters for milling and drilling operations on the energy demand (Mori et al. 2011) and the modeling of the specific energy consumption (SEC), which describes the energy consumption of the machine tool for removing 1 cm^3 workpiece material (Kara and Li 2011). Within this approach, the SEC was modeled depending on the material removal rate and for wet and dry machining conditions. Li et al. (2012) further extended this approach by the consideration of resource consumption and surface quality of grinding processes, while Li and Kara (2011) applied the approach to turning processes.

A variety of other approaches also focused on the energy demand of turning processes. Zhong et al. (2016) experimentally evaluated, among others, the accuracy of different models proposed in literature to calculate the energy demand. Velchev et al. (2014) modeled the direct energy consumption influenced by process parameters and optimized them. Further, Mativenga and Rajemi (2011) optimized the combination of process parameters regarding tool life and energy demand of the machine tool, while Guo et al. (2012) proposed a two-step approach to determine the optimal process parameters considering a specified surface roughness and minimal energy consumption. Based on the selection of process parameters, a multi-objective approach to improve various indicators such as carbon emissions, energy demand, surface roughness or costs simultaneously was proposed by different studies (Jin et al. 2009; Robiglio 2017; Sihag and Sangwan 2018).

With regard to milling processes, other approaches such as Liu et al. (2015), also focused on modeling the energy consumption in relation to the process parameters. In addition to the direct energy consumption of the milling process, Chen et al. (2018), Liu et al. (2016), Rief (2012) also considered the embodied energy of auxiliary materials such as tools or cutting fluid to optimize the process.

A multi-objective optimization approach, which considers not only the energy consumption, but also other indicators such as surface quality, tool life or costs was developed by a variety of authors (Kuram et al. 2013; Li et al. 2017; Wang et al. 2014; Yan and Li 2013). However, the approaches differ with regard to the used empirical models and optimization algorithm. Besides the cutting velocity, feed rate and depth of cut, Kuram et al. (2013) also focused on the use of different cutting fluids as influencing process parameters.

Peripheral system level

Within the derived scientific network, research approaches at this level are cited less frequently on average. A clear outlier, however, is the study by Pusavec et al. (2010) with more than 200 citations. Within this study, the cutting fluid strategies flooding, cryogenic and high pressure jet-assisted machining were analyzed for a turning process from an environmental perspective using a LCA. Fratila (2010) also performed a LCA study to compare the environmental performance of the strategies flooding and near-dry machining for a milling process. Overall, both studies have demonstrated a higher environmental impact of flood machining with conventional cutting fluids in comparison to the other strategies of the respective study. Further, Clarens et al. (2008) also compared the environmental impacts of different cutting fluids and application strategies within a LCA, demonstrating the environmental benefits of a gas-based against conventional system. In contrast, Priarone et al. (2016) analyzed and modeled the impact of the cutting fluid strategies dry, MQL and flooding from a technological perspective with a focus on tool life and energy demand of the machining system. This study revealed that the flooding strategy leads to a reduced energy consumption at high cutting speeds and MRR. Different cutting fluid strategies have also been investigated by several authors regarding the technological impact within a turning process (Deiab et al. 2014; Faga et al. 2017; Shokoohi et al. 2015).

Besides the influence of different application strategies, the influence of different types of cutting fluids were also investigated and compared with regard to their performance. Kuram et al. (2010) analyzed the resulting surface roughness and thrust force using various vegetable based and mineral oil based cutting fluids, while Soković and Mijanović (2001) compared semi-synthetic fluids and assessed the torque depending on the cutting speed. Different cutting fluids were also investigated and evaluated by Axinte et al. (2003) with regard to their technological performance within different machining processes. Winter et al. (2015) also investigated the resulting surface roughness from the use of different water-miscible and non-water miscible cutting fluids for a grinding process. In addition, the cutting power, costs and GWP were also examined within this approach. The approach presented by Dettmer (2006) focuses on non-water miscible fluids, but provides a comprehensive environmental assessment using a LCA, an economic assessment using LCC and a technological assessment considering surface quality, process forces and grinding wheel wear. A qualitative comparison of the technological, economic and environmental performance of different strategies and cutting fluids was presented by Benedicto et al. (2017), while Sutherland et al. (1997) proposed a model-based assessment of their cooling performance, costs, fluid loss and mist generation as well as health hazard.

Eckebrecht (2000) presented the technological and economic assessment of cutting fluids and strategies with a focus on energy consumption and cutting fluid filtration systems.

The modeling and improvement of the energy demand of cutting fluid system elements such as pumps or cooling was also the focus of different studies (Triebs 2015; Weber et al. 2016; Züst 2017). Further studies also investigated the generation of cutting fluid emissions and fluid loss. Bell et al. (1999) modeled three primary mechanisms for the generation of cutting fluid emissions - evaporation, splash and spin-off. In addition to the mechanisms splash and spin-off, Chen et al. (2000) also focused on the particle size distribution as well as the impact of process parameters and fluid properties. The approach presented by Petuelli (2002) focused on regression models for the description of drag-out due to emissions but also due to wetting of workpieces, chips and filtration materials.

Machining system level

At the level of machining systems, it can be noted that some earlier studies from the years 2003–2009 have a significant scientific impact and were cited extensively within the derived network. In contrast, various dissertations from more recent years cite a large number of other studies within this network, but are rarely cited themselves. These approaches often consider not only individual manufacturing processes, but also a process chain consisting of different processes.

The approach proposed by Larek (2012) focused on the energy and resource consumption of process chains based on empirical measurements, while the approach developed by Schlosser (2013) focused on predicting the economic and environmental performance using a LCA. Regarding the design of process chains, Eisele (2014) and Schrems (2014) mainly considered the energy consumption of different processes and components. However, while Schrems aimed to estimate the energy consumption during the planning phase, Eisele aimed to identify and evaluate energy improvement measures. The measurement and evaluation of energy efficiency of general machine tools as well as improvement measures were also in the focus of the approach by Schudeleit (2017), Anderberg (2012) and Zein (2012). Further, the energy and resource flows within machine tool systems were analyzed by Schiefer (2001) from an environmental perspective and modeled by Kuhrke (2011) for different subsystems using data sheets and reference measurements.

Various other research approaches also developed a general and process independent method. Xue et al. (2007) modeled the relation between input and output flows of general manufacturing systems with the aim to balance them, while Dietmair and Verl (2009) presented a generic method to model energy consumption related to operational state including stochastic simulations. The development of a sustainable manufacturing assessment framework including social, economic and environmental aspects to compare and rank different alternatives was presented by Zhang and Haapala (2015). In order to further support decision making, different studies also focused on the development of software-tools. Avram et al. (2011) presented a multi-criteria decision-making approach considering technological, economic and environmental aspects using the analytical hierarchy process, while Jiang et al. (2012)

focused on the environmental performance based on input and output flows of 24 different manufacturing processes e.g. tuning, milling and grinding. Gutowski et al. (2006) presented an empirical modeling approach to describe the specific energy demand related to process rate for various manufacturing processes. A rather descriptive approach was proposed by Dahmus and Gutowski (2004) on a system-level analysis of the energy demand based on measured load profiles of milling machines. With regard to milling machines, Draganescu et al. (2003) empirically modeled the machine tool efficiency and specific consumed energy as a function of process parameters, Devoldere et al. assessed the energy improvement potential depending on operational state and non-productive time and Altıntaş et al. (2016) developed models to estimate the theoretical energy consumption and aimed to reduce it by optimizing the process parameters. Besides modeling energy consumption and costs of milling processes, the research approach by Diaz-Elsayed (2013) also focused on production scheduling to increase energy efficiency. With regard to turning machines, Guo et al. (2015) proposed an operation-mode based method to predict the energy consumption by material removal simulation and optimized process parameters and toolpath strategies to reduce energy consumption. Further, Priarone et al. (2018) assessed the environmental and economic impacts of wet and dry turning strategies on the entire system. Winter (2016), however, focused on grinding and presented an approach to describe, model, evaluate and improve the eco-efficiency of grinding processes and systems. The technological and environmental trade-off between grinding and hard cutting were analyzed by Kundrák et al. (2006).

Out of the briefly presented research overview involving 70 approaches, 17 studies are selected over the discussed focus levels. These approaches comprise the three central perspectives by consideration of the machining systems and relevant subsystems as well as the environmental and economical dimension in addition to the technological dimension. They also demonstrate an innovative approach from a methodical perspective, which goes beyond a literature review. Moreover, the approaches are selected according to a high scientific impact based on the total number of citations or a high relative impact within the derived citation network. Following the procedure shown in Fig. 3.1, the selected approaches are further examined and evaluated based on defined criteria in order to derive a research gap.

3.2 Criteria for the Comparative Evaluation of Research Approaches

For a structured analysis and comparative assessment of the selected approaches, the definition of specific criteria is required. These criteria should be independent and comprehensively cover all relevant aspects of the research topic. Therefore, the defined criteria are divided into four main areas with 17 specified sub-criteria. An overview of the classification is presented in Fig. 3.3 and described in more detail in the following. However, as such an assessment is always subjective, it is important

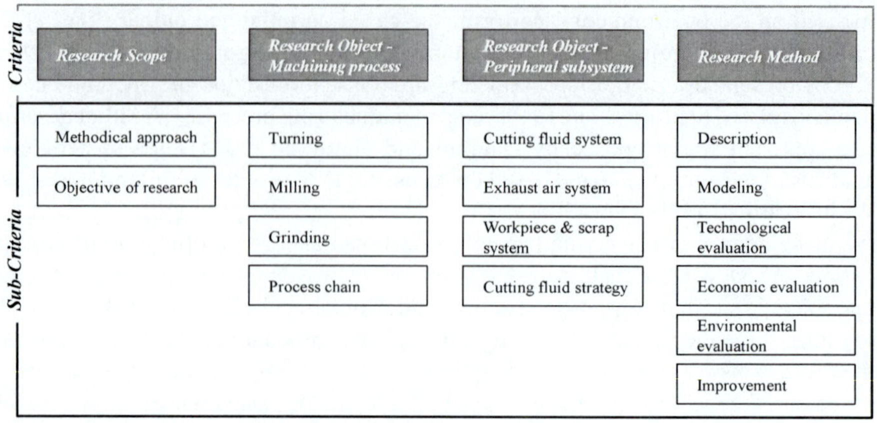

Fig. 3.3 Overview of selected criteria and sub-criteria

that the criteria are also as comprehensible and transparent as possible. For this purpose, each criterion is further classified into four content-related gradations with specified characteristics, which are illustrated in Table 3.1. For each characteristic fulfilled, an approach is evaluated by a quarter (◔). Thus, up to (●) can be achieved depending on the degree of fulfillment.

Research scope

This criterion research scope is selected in order to analyze and differentiate the selected *methodical approach* and examined *objective of research*. The methodological approach distinguishes whether a theoretical description is conducted, a method or also a software tool is being developed or experiments are carried out. The research objective can also be evaluated according to four characteristics. It focuses on the objective to create general knowledge, to provide a decision support regarding the machining process (e.g. selection of process parameters), regarding the peripheral design (e.g. dimension of cutting fluid pump) or also regarding the configuration of a process chain.

Research object

This criterion focuses on the elements of the machining system that are considered and examined within each research approach. According to the structure described in the previous chapter, it is divided into eight sub-criteria and differs mainly between the *machining processes* and the elements of the *peripheral subsystem*. The machining processes involve processes with geometrically defined cutting edges—*turning* and *milling* as well as processes with geometrically undefined cutting edges—*grinding*. All processes are evaluated whether they are considered within a general approach, described theoretically, analyzed by simple metering or by an extensive design of

Table 3.1 Detailed overview of selected criteria and sub-criteria

Degree of criteria fulfillment	○	+ ◔	+ ◔	+ ◔	+ ◔
Research scope					
Methodical approach	Not specified	Theoretical description	Method development	Software/tool development	Experimental research
Objective of research	Not specified	General knowledge	Decision support- process design	Decision support- peripheral design	Decision support- Process chain
Research object - machining process					
Turning Milling Grinding	Not considered	Generic approach	Theoretical	Simple metering	Design of experiments
Process chain	Not considered	Multiple processes	Variation of processes	Variation of products	Variation of program
Research object - peripheral subsystem					
Cutting fluid system	Not considered	Black box	Pump	Filtration	Supply (e.g. nozzle)
Exhaust air system	Not considered	Black box	Mist and vapor	Filtration	Hardware
Workpiece and scrap treatment system	Not considered	Black box	Cleaning	Chip transport	Scrap de-oiling
Application strategy	No variation	Variation of cutting fluid supply	Alternative cutting fluid formulation	Alternative strategy	Centralized/decentralized system
Research method					
Description	Not considered	Technical	Costs	Energy flows	Resource flows

(continued)

Table 3.1 (continued)

Degree of criteria fulfillment	○	+ ◔	+ ◔	+ ◔	+ ◔
Modeling	Not considered	System identification	Empirical modeling	Physical modeling	Simulation
Technological evaluation			Quantitative assessment of:		
	Not considered	Qualitative assessment	– one factor	– multiple factors	– complete process
Economic evaluation	Not considered	Qualitative assessment	– costs	– a basic cost model	– a complete cost model
Environmental evaluation	Not considered	Qualitative assessment	– one factor	– multiple impact factors	– a full LCA
Improvement	Not considered	Technical measures	Organizational measures	Mathematical optimization	Multi-objective optimization

experiments. The distinction between simple metering and performance of experiments is relevant because it is, for example, common to measure a load profile to identify main electrical consumers, but a conclusion about the impact of the parameters on the energy demand can only be achieved by experiments with detailed parameter variations. Additionally, the combination of multiple machining processes, the variation of different processes and different products as well as the variation of production program are considered and evaluated within the sub-criterion *process chain*. However, the elements of the peripheral subsystem such as the *cutting fluid system*, the *exhaust air system* or the *workpiece and scrap treatment system* are also considered as relevant research objects and are therefore represented in sub-criteria. Each of these subsystems are evaluated according to their components considered, which are either critical in terms of electrical consumption e.g. cutting fluid pump and fan, resource consumption e.g. scrap de-oiling and emissions or functional performance e.g. filtration. Additionally, the sub-criterion *application strategy* further focuses on the variety of different cutting fluid strategies, which have a strong impact on the entire machining system. It comprises the variation of cutting fluid supply e.g. by different flow rates, the use of alternative cutting fluid formulations e.g. by using mineral oil free formulations or comparing water-based with non-water-based fluids or the investigation of alternative strategies e.g. MQL or dry machining. Further, the use of centralized and decentralized cutting fluid as well as exhaust air systems are also considered.

Research method

This criterion describes whether and to which extent different methods are used within each research approach and is therefore divided into six sub-criteria, which comprise the methods. As a basic method, it is therefore examined if a *description* of the technical context, the general costs or the energy and resources flows is carried out. These are the relevant aspects and flows from a holistic approach that includes an economic, environmental and technological perspective. At the next method level, it is also evaluated whether a *modeling* of the interdependencies is performed. This may include a general system identification with the examination of input and output flows of a specific system or sub-system. Further, this could also include an empirical modeling approach with e.g. regressions models based on experimental data or a physical modeling approach. The implementation of a simulation based on the derived models may also be included and assessed within this criterion. The level of evaluation methods is divided into the three groups and sub-criteria. According to the holistic perspective, these include the *technological*, *economic* and *environmental evaluation*. All three sub-criteria are assessed in a similar manner, where a distinction is made, whether a simple qualitative assessment with a description of the issues and results is performed or a quantitative assessment with the actual calculation of the issues and results. The quantitative assessment is further differentiated by the calculation of a single factor, multiple factors or the complete results e.g. LCA or cost model. The sub-criteria *improvement* distinguishes between technical measures such as the replacement of components as well as organizational measures such as an improved planning of process times. Further, this sub-criteria considers the mathematical optimization of different parameters regarding one objective or a multi-objective optimization regarding multiple objectives.

3.3 Evaluation of Selected Approaches

The selected research approaches are further analyzed regarding the fulfillment of the defined set of criteria from Sect. 3.2 and the results are summarized in Table 3.2 within this section. The approaches are presented according to the same order for the three focus levels process, peripheral and machining system.

Table 3.2 Comparative evaluation of selected research approaches

Focus level	*Process*					*Peripheral System*					*Machining System*						
Research approach	Munoz & Sheng 1995	Sirinvasan & Sheng 1999	Mori et al. 2011	Li et al. 2012	Chen et al. 2018	Eckebrecht 2000	Dettmer 2006	Clarens et al. 2008	Pusavec et al. 2010	Priarone et al. 2016	Dahmus & Gutowski 2004	Avram et al. 2012	Larek 2012	Schlosser 2013	Eisele 2014	Schrems 2014	Winter 2016
Research scope																	
Methodical approach	◑	◕	◓	◕	◕	◓	◕	◑	◔	◕	◔	◕	●	●	●	◕	●
Objective of research	◑	◕	◑	◕	◑	◑	◑	◑	◑	◕	◔	◕	●	◕	◑	●	◕
Research object - Machining process																	
Turning	◑	◑	○	○	○	◑	◔	◔	◑	◕	◔	◔	●	◕	◑	◕	◔
Milling	●	●	●	○	●	◑	◔	◔	◔	◔	◑	◔	◔	◕	◕	◔	◔
Grinding	○	○	○	●	○	◕	◕	◔	◔	○	○	◔	●	○	○	◔	●
Process chain	○	◔	◔	○	○	◔	○	○	○	○	○	○	◕	◕	◒	◕	○
Research object - Peripheral subsystem																	
CF system	◔	◔	○	◑	◔	◕	◔	◑	◑	◑	◕	◑	◔	◔	◕	●	◕
Exhaust air system	◔	◔	○	◔	○	◔	○	◔	○	○	○	◓	◑	◔	◓	◔	◑
WP & scrap system	◔	○	○	○	○	◕	●	○	○	○	◔	◑	◔	◔	○	◑	◔
Application strategy	◔	◔	○	◔	◔	●	◒	●	◕	◒	○	◐	○	◓	◔	◔	◑
Research methods																	
Description	◕	◕	◑	◕	◕	●	●	◐	◐	◕	◐	●	◕	●	◒	◕	●
Modelling	◕	◕	◔	◑	◑	◔	◑	◔	◑	◑	◔	◑	◕	◕	●	●	◕
Techn. evaluation	◕	◕	◔	◕	◑	◑	◑	○	○	◑	○	◔	◑	◕	◔	○	●
Economic evaluation	○	○	○	○	○	◑	◕	○	○	○	○	◔	◑	●	◕	◑	●
Environ. evaluation	◑	◕	◑	◑	◑	◑	●	●	●	◑	◔	◔	◑	●	◑	◑	●
Improvement	◔	●	◔	◔	◕	◑	◔	◔	◔	◑	○	◑	●	◑	●	◑	●

With regard to the process level, Munoz and Sheng (1995) presented a model based method to assess the environmental impact of machining processes and to support the design of process parameters. The approach focuses on machining processes with geometrically defined cutting edges and is applied within a milling case study. Although the peripheral subsystems are considered rather generally, the cutting fluid loss e.g. due to evaporation is modeled in more detail. However, a variation regarding the application strategy and design of the subsystems are not further considered within this approach. Besides the technological evaluation of tool wear and cutting force, the environmental impact of waste streams and the influence of technical measures is also evaluated. An economic evaluation, however, is not included.

Srinivasan and Sheng (1999) presented an approach which is based on the approach by Munoz and Sheng (1995) regarding process modeling, but focuses on the support of process planning decisions for the selection of cutting fluid, tools and process paths. Within this study, a variation of turning and especially milling processes are theoretically and experimentally considered. However, as in the previous study, the peripheral subsystems are considered more generally, but the environmental impact of alternative cutting fluid formulations is assessed in this study as well. Similarly, the assessment is limited to the technological and environmental impact, but not to costs.

The approach by Mori et al. (2011) focused on technical measures to reduce power consumption of machining processes. In the context of this study, reductions were achieved by synchronizing the spindle with the feed system and selecting the process parameters to shorten the process time without compromising technological aspects. These measures were experimentally analyzed for different milling and drilling operations, but without explicitly considering the peripheral subsystem or cutting fluid strategy.

The study presented by Li et al. (2012) is based on the empirical approach proposed by Kara and Li (2011) regarding the relationship between energy consumption and process parameters, which has also been applied for turning processes (Li and Kara 2011). As part of this study, the approach was applied to a grinding process and extended by considering the quality performance and the impact of peripheral subsystems to support the process. The environmental analysis used the functional unit of grinding 1 kg of the material 100Cr6, while achieving a given surface roughness value R_z. The results are expressed by CO_2 emissions and demonstrate a high impact due to the machining process and the cutting fluid pump. The evaluation of costs was not part of this study and possible improvements were limited to the investigation of process parameters.

Chen et al. (2018) proposed a modeling approach that considered the direct energy consumption by the machine tool and indirect energy consumption by the embodied energy of cutting fluids and cutting tools. Further, a multi-objective optimization model was used and solved by particle swarm algorithm to improve the energy and machining efficiency. The approach was experimentally verified for a milling process using cutting velocity, cutting depth and feed rate as optimization variables. The experimental results demonstrated that the cutting velocity has the highest impact and that a maximized machining efficiency does not also result in a maximized energy efficiency, which shows the need for a multi-objective method. However, the presented approach mainly focused on the milling process itself and the peripheral subsystems were only considered by the embodied energy of cutting fluids. Further, an economic evaluation and other machining processes were not analyzed.

Regarding research approaches at the level of peripheral subsystems, the study presented by Eckebrecht (2000) aimed at an environmentally compatible design of manufacturing processes. Apart from concepts for the knowledge transfer into companies, the main focus was on measures to reduce the environmental impact. They were applied to different machining processes with defined and undefined cutting edges and verified by basic measurements. The measures involved the use of alternative cutting fluid strategies such as MQL, the de-oiling of grinding swarf to reuse fluid and material as well as reducing the use of energy for grinding processes through e.g. a more efficient use of the pump and cooling system. However, the application of MQL revealed that it may also lead to a negative impact on costs and environment due to high energy demand for the compressed air and increased tool wear. Overall, the general focus was more on the peripheral subsystems and neglected the impact of process parameters.

The study presented by Dettmer (2006) focused on non-water-miscible cutting fluids and analyzed the technological, economic and environmental impact of animal fat and used cooking oil esters in comparison to a mineral oil based product. The cutting fluids were used and analyzed for grinding processes and examined with regard to scrap treatment and the cutting fluid system but without the exhaust air system. The conducted investigations demonstrated that the mineral oil based cutting fluid has the highest environmental impact but also the lowest price. However, due to lower refilling volumes, the other cutting fluids could compensate for this after a relatively short time. The impact of supply strategies or improvement of process parameters were not analyzed.

Clarens et al. (2008) focused on the environmental comparison of alternative cutting fluid strategies. These involved water-based cutting fluids using mineral oil and rapeseed oil as well as MQL and supercritical CO_2 using rapeseed oil. The functional unit was defined as service of each cutting fluid strategy to one machine tool in a decentralized system for one year. For each strategy, the analysis considered the production, end-of-life and use phase with a focus on e.g. drag-out, pumps and flow rates. Overall, the authors concluded that using gas-based instead of water-based cutting fluid strategies could be advantageous from an environmental perspective. However, the economic and technological performance of the cutting fluid strategies, which could also have an impact on the results due to e.g. tool wear or process parameters, were not taken into account within this study.

Pusavec et al. (2010) presented a study on alternative cutting fluid strategies as well as focusing on cryogenic and high pressure jet assisted machining (HPJAM). In the context of the study, a comparative LCA was conducted to assess the environmental performance, however, also assuming that the machining process quality does not vary depending on the cutting fluid strategy. Thus, the influence of different technological performances was not taken into account. The conducted LCA focused on the production and supply of the cutting fluid and further excluded the exhaust air as well as workpiece and scrap treatment system. The results demonstrated that the cryogenic strategy can lead to reduced solid waste, acidification, global warming potential and water usage, but increased energy use compared to mineral oil based cutting fluids.

The study presented by Priarone et al. (2016) compared different cutting fluid strategies as well, but also focused on the technological performance within a turning process. The study analyzed the direct energy of the machine tool and indirect energy related to the use of cutting fluids and tools. This also involved the energy to produce and supply the cutting fluids, but not the exhaust air and scrap treatment system. For each cutting fluid strategy - dry, MQL and flooding - different process parameters were experimentally investigated regarding the specific energy consumption and thus, optimal parameters were identified. It was determined that wet turning could minimize the total energy consumption by higher possible cutting speeds, whereas dry turning requires a lower MRR due to increasing tool wear. Furthermore, the study revealed that MQL is not an advantageous alternative due to the high compressed air demand. The approach, however, is limited to energy and does not take all subsystems into account that are affected by the choice of strategy.

At machining system level, Dahmus and Gutowski (2004) presented an environmental analysis of machining including material and cutting fluid preparation as well as cleaning. The focus of the study was on machining processes with geometrically defined cutting edges and their energy consumption. The energy breakdown analysis of different machine tools demonstrated that the actual cutting energy is rather small compared to the total energy demand and that in some cases the embodied energy of the workpiece material exceeds the energy demand of the machine tool. This analysis is a rather descriptive approach without further considering process parameter or quantitative assessments of the technological, economic or environmental perspective.

The approach by Avram et al. (2011) presented a method and software development to support design and operation of machining systems. This involved material removal processes and peripheral subsystems such as chip removal and cutting fluid systems to support them. Depending on the use case, technological, economic and environmental criteria were evaluated at process and system level for alternative strategies regarding cutting fluids, tools and parameters. For this evaluation, the software used a database with experimentally collected data, which had to be performed for each scenario. The alternatives were further structured and ranked by using the analytic hierarchy process (AHP), which supported the user to prioritize different criteria with a pairwise comparison. This approach was applied to a milling process within the study and concluded that MQL was the preferred alternative compared to wet and dry machining. This approach, however, is rather descriptive and does not quantify specific indicators. Further, the application of the approach requires a comparatively high effort to carry out the required experiments.

The study by Larek (2012) aimed to develop an approach, which does not require experimental data to estimate the consumption of energy and resources of manufacturing processes but also process chains. These involve machining processes such as turning and grinding as well as hardening processes. By using a genetic optimization algorithm, process parameters should be selected in order to reduce energy and resource consumption as well as costs. This approach included the use of cutting fluids, but overall peripheral subsystems and their impact are not considered in detail.

Schlosser (2013) presented an approach for the prediction of the energy and resource consumption of machining processes in order to support sustainable decision making regarding the selection of process parameters, machine tools and process chains. Within a software tool, different production alternatives were to be evaluated not only from a technological and economic perspective, but also from an environmental perspective with a focus on existing machining systems. The correlations and processes were described mathematically and partly verified experimentally. The models, however, focused on processes with geometrically defined cutting edges and are not transferable to processes with undefined cutting edges such as grinding. Although the approach considered the use of cutting fluids, different strategies and their impacts on the overall system were not considered.

An approach to simulate and improve the energy consumption of machining systems was presented by Eisele (2014). Energy-relevant components and systems of a machine tool were modeled and implemented into a simulation tool. These included, for example, the cutting fluid and exhaust air system but ignoring the impact of alternative cutting fluid strategies. The improvement focused on the use of more efficient components, a demand-oriented control and standby shutdown. The improvement measures did not consider the process parameters and technological impact. Further, the approach focused only on machining processes with geometrically defined cutting edges and energy consumption with the result that, for example, grinding processes and the influence of resources are not taken into account.

Schrems (2014) developed another approach to support the selection of alternative production resources by a model-based estimation of the energy consumption. This focused on individual machining systems, but also on a variety of systems within a process chain, which were modeled using the simulation tool Plant Simulation®. Besides machining systems, the approach also included hardening and cleaning systems. For each subsystem, relevant components were modeled such as the pump or the nozzle for the cutting fluid system. However, the impact of alternative cutting fluid strategies was not considered within the approach. Furthermore, the selection of different system configurations was mainly based on energy consumption, without considering technological aspects.

An approach to increase the eco-efficiency of existing and new grinding systems was presented by Winter (2016). Based on physical and empirical models, the technological, economic and environmental impacts of different grinding process parameters and system configurations were calculated and improved by a multi-objective approach. Within this approach, the focus was on chip formation, grinding and dressing processes as well as peripheral subsystems such exhaust air and cutting fluid system. Other subsystems such as cleaning machines or scrap processing were not considered. Moreover, the focus was explicitly on grinding, meaning that other processes with geometrically defined cutting edges and also process chains were not taken into account.

3.4 Derivation of Research Demand

The results of the comparative evaluation are presented in Table 3.2. Overall, it can be concluded that most criteria are already partly considered by different approaches. There is no approach that includes all relevant criteria for a technological, environmental and economic planning and operation of machining systems. Moreover, the degree of fulfilment is rather different for each criterion and highly depends on the respective focus level.

At process level, a research gap could be identified with regard to the investigation of the economic impact. The reviewed approaches mainly focused on the interactions between process parameters and technological as well as environmental impacts, while none of them considered the economic impact. Also with regard to the influence of the peripheral subsystems, the interactions were not examined in detail. In contrast, approaches at the peripheral subsystem level focused more on the interactions between subsystems and environmental impacts and rather less on technological and economic impacts. Most of these approaches examined in particular the influence of different cutting fluid strategies. For example, several authors have compared the environmental impact of MQL and wet machining and have concluded with different results (Clarens et al. 2008; Eckebrecht 2000; Priarone et al. 2016). While Eckebrecht and Priarone et al. did not determine an environmental advantage by using MQL, Clarens et al. determined the opposite results for several impact categories. This is mainly due to an inconsistent approach, which in all cases did not consider the entire system affected by the respective cutting fluid strategy. The reviewed studies at machining system level showed a more comprehensive, but also more general approach. In several cases, for example, experiments were mainly carried out to verify results by simple metering of the load profile and not to investigate the influence of process parameters.

Additional findings were also identified, which are generally valid for the reviewed approaches regardless of the focus level. While each approach investigated the energy consumption, the resource consumption was not always part of the investigation and in several cases only represented by the embodied energy of e.g. cutting fluids and tools (Chen et al. 2018; Dahmus and Gutowski 2004). Further, the reviewed approaches either focused on machining processes with geometrically defined cutting edges such as milling or turning or on processes with undefined cutting edges such as grinding. Although some approaches are generally transferable to other processes, only Eckebrecht (2000) and Larek (2012) actually considered grinding as well as turning/milling processes. This differentiated consideration is comprehensible, as the technological modeling of the processes differs significantly. However, for an investigation of process chains with different processes a comprehensive consideration is required. With regard to the defined criteria for this work, a further significant gap could be identified for all examined approaches regarding the impact of cutting fluid strategies on the process chain. A possible reduction or elimination of cleaning processes between different processes along the process chain depending on the strategy was not investigated by any approach. A distinction between centralized or decentralized cutting fluid as well as exhaust air systems was also not the object of almost any research approach. However, with regard to e.g. pressure losses due to piping or capacity planning this could be particularly interesting for the design of efficient process chains.

In summary, it can be concluded from this literature review that there is still a significant research demand on various aspects:

- Simultaneous examination of the technical, environmental and economic perspectives for the research objects.
- Comprehensive consideration of all subsystems and components that are affected by the use of different cutting fluid strategies.
- Model-based description of the effects of varying process parameters at machining system level and in combination with cutting fluid strategies.
- Consideration of relevant energy as well as resource flows throughout the entire machining system including subsequent subsystems.
- Integration of multiple machining processes with defined and undefined cutting edges to investigate the effects of cutting fluid strategies on process chains.

References

Altıntaş RS, Kahya M, Ünver HÖ (2016) Modelling and optimization of energy consumption for feature based milling. Int J Adv Manuf Technol 86:3345–3363. https://doi.org/10.1007/s00170-016-8441-7

Anderberg S (2012) Methods for improving performance of process planning for CNC machining. PhD thesis, Chalmers University of Technology, Göteborg, Sweden

Avram O, Stroud I, Xirouchakis P (2011) A multi-criteria decision method for sustainability assessment of the use phase of machine tool systems. Int J Adv Manuf Technol 53:811–828. https://doi.org/10.1007/s00170-010-2873-2

Axinte D, Axinte M, Tannock JDT (2003) A multicriteria model for cutting fluid evaluation. Proc Inst Mech Eng Part B: J Eng Manuf 217:1341–1353. https://doi.org/10.1243/095440503322617117

Bell DD, Chou J, Nowag L, Liang SY (1999) Modeling of the environmental effect of cutting fluid. Tribol Trans 42:168–173. https://doi.org/10.1080/10402009908982204

Benedicto E, Carou D, Rubio EM (2017) Technical, economic and environmental review of the lubrication/cooling systems used in machining processes. Adv Mater Process Technol Conf Procedia Eng 184:99–116. https://doi.org/10.1016/j.proeng.2017.04.075

Chen Z, Atmadi A, Stephenson DA, Liang SY, Patri KV (2000) Analysis of cutting fluid aerosol generation for environmentally responsible machining. CIRP Ann 49:53–56. https://doi.org/10.1016/S0007-8506(07)62894-7

Chen X, Li C, Jin Y, Li L (2018) Optimization of cutting parameters with a sustainable consideration of electrical energy and embodied energy of materials. Int J Adv Manuf Technol 96:775–788. https://doi.org/10.1007/s00170-018-1647-0

Clarens AF, Zimmerman JB, Keoleian GA, Hayes KF, Skerlos SJ (2008) Comparison of life cycle emissions and energy consumption for environmentally adapted metalworking fluid systems. Environ Sci Technol 42:8534–8540. https://doi.org/10.1021/es800791z

Dahmus JB, Gutowski TG (2004) An environmental analysis of machining. In: Proceedings of ASME international mechanical engineering congress and R&D exposition, pp 643–652. https://doi.org/10.1115/IMECE2004-62600

Deiab I, Raza SW, Pervaiz S (2014) Analysis of lubrication strategies for sustainable machining during turning of titanium Ti-6Al-4V alloy. In: 47th CIRP conference on manufacturing systems, vol 17, pp 766–771. https://doi.org/10.1016/j.procir.2014.01.112

Dettmer T (2006) Nichtwassermischbare Kühlschmierstoffe auf Basis nachwachsender Rohstoffe. Dr.-Ing. dissertation, Technische Universität Braunschweig. Vulkan Verlag, Essen, Germany

Devoldere T, Dewulf W, Deprez W, Willems B, Duflou JR. Improvement potential for energy consumption in discrete part production machines. In: 14th CIRP international conference on life cycle engineering, Tokyo, Japan, pp 311–316. https://doi.org/10.1007/978-1-84628-935-4_54

Diaz-Elsayed N (2013) Development of energy models for production systems and processes to inform environmentally benign decision-making. PhD thesis, University of California, Berkeley, USA

Dietmair A, Verl A (2009) A generic energy consumption model for decision making and energy efficiency optimisation in manufacturing. Int J Sustain Eng 2:123–133. https://doi.org/10.1080/19397030902947041

Draganescu F, Gheorghe M, Doicin CV (2003) Models of machine tool efficiency and specific consumed energy. J Mater Process Technol 141:9–15. https://doi.org/10.1016/S0924-0136(02)00930-5

Eckebrecht J (2000) Umweltverträgliche Gestaltung von spanenden Fertigungsprozessen. Dr.-Ing. dissertation, Universität Bremen. Shaker Verlag, Aachen, Germany

Eisele C (2014) Simulationsgestützte Optimierung des elektrischen Energiebedarfs spanender Werkzeugmaschinen. Dr.-Ing. dissertation, TU Darmstadt. Shaker Verlag, Aachen, Germany

Faga MG, Priarone PC, Robiglio M, Settineri L, Tebaldo V (2017) Technological and sustainability implications of dry, near-dry, and wet turning of Ti-6Al-4V alloy. Int J Precis Eng Manuf Green Technol 4:129–139. https://doi.org/10.1007/s40684-017-0016-z

Fratila D (2010) Macro-level environmental comparison of near-dry machining and flood machining. J Clean Prod 18:1031–1039. https://doi.org/10.1016/j.jclepro.2010.01.017

Guo Y, Loenders J, Duflou J, Lauwers B (2012) Optimization of energy consumption and surface quality in finish turning. In: 5th CIRP conference on high performance cutting, vol 1, pp 512–517. https://doi.org/10.1016/j.procir.2012.04.091

Guo Y, Duflou JR, Qian J, Tang H, Lauwers B (2015) An operation-mode based simulation approach to enhance the energy conservation of machine tools. J Clean Prod 101:348–359. https://doi.org/10.1016/j.jclepro.2015.03.097

Gutowski TG, Dahmus JB, Thiriez A (2006) Electrical energy requirements for manufacturing processes. In: 13th CIRP international conference on life cycle engineering, Leuven, Belgium, pp 623–627

Jiang Z, Zhang H, Sutherland JW (2012) Development of an environmental performance assessment method for manufacturing process plans. Int J Adv Manuf Technol 58:783–790. https://doi.org/10.1007/s00170-011-3410-7

Jin K, Zhang HC, Balasubramaniam P, Nage S (2009) A multiple objective optimization model for environmental benign process planning. In: 16th International conference on industrial engineering and engineering management, pp 869–873. https://doi.org/10.1109/ICIEEM.2009.5344194

Kara S, Li W (2011) Unit process energy consumption models for material removal processes. CIRP Ann 60:37–40. https://doi.org/10.1016/j.cirp.2011.03.018

Kuhrke B (2011) Methode zur Energie- und Medienbedarfsbewertung spanender Werkzeugmaschinen. Dr.-Ing. dissertation, TU Darmstadt. epubli, Berlin, Germany

Kundrák J, Mamalis AG, Gyáni K, Markopoulos A (2006) Environmentally friendly precision machining. Mater Manuf Process 21:29–37. https://doi.org/10.1080/AMP-200060612

Kuram E, Ozcelik B, Demirbas E, Simsek E (2010) Effects of the cutting fluid types and cutting parameters on surface roughness and thrust force. The World Congress on Engineering, London

Kuram E, Ozcelik B, Bayramoglu M, Demirbas E, Simsek E (2013) Optimization of cutting fluids and cutting parameters during end milling by using D-optimal design of experiments. J Clean Prod 42:159–166. https://doi.org/10.1016/j.jclepro.2012.11.003

Kwon Y, Fischer GW (2003) A novel approach to quantifying tool wear and tool life measurements for optimal tool management. Int J Mach Tools Manuf 43:359–368. https://doi.org/10.1016/S0890-6955(02)00271-7

Larek R (2012) Ressourceneffiziente Auslegung von fertigungstechnischen Prozessketten durch Simulation und numerische Optimierung. Dr.-Ing. dissertation, Universität Bremen. Shaker Verlag, Aachen, Germany

Li W, Kara S (2011) An empirical model for predicting energy consumption of manufacturing processes: a case of turning process. Proc Inst Mech Eng Part B: J Eng Manuf 225:1636–1646. https://doi.org/10.1177/2041297511398541

Li W, Winter M, Kara S, Herrmann C (2012) Eco-efficiency of manufacturing processes: a grinding case. CIRP Ann 61:59–62. https://doi.org/10.1016/j.cirp.2012.03.029

Li C, Chen X, Tang Y, Li L (2017) Selection of optimum parameters in multi-pass face milling for maximum energy efficiency and minimum production cost. J Clean Prod 140:1805–1818. https://doi.org/10.1016/j.jclepro.2016.07.086

Liu N, Zhang YF, Lu WF (2015) A hybrid approach to energy consumption modelling based on cutting power: a milling case. J Clean Prod 104:264–272. https://doi.org/10.1016/j.jclepro.2015.05.049

Liu ZY, Guo YB, Sealy MP, Liu ZQ (2016) Energy consumption and process sustainability of hard milling with tool wear progression. J Mater Process Technol 229:305–312

Mativenga PT, Rajemi MF (2011) Calculation of optimum cutting parameters based on minimum energy footprint. CIRP Ann 60:149–152. https://doi.org/10.1016/j.cirp.2011.03.088

Mori M, Fujishima M, Inamasu Y, Oda Y (2011) A study on energy efficiency improvement for machine tools. CIRP Ann 60:145–148. https://doi.org/10.1016/j.cirp.2011.03.099

Munoz AA, Sheng P (1995) An analytical approach for determining the environmental impact of machining processes. J Mater Process Technol 53:736–758. https://doi.org/10.1016/0924-0136(94)01764-R

Petuelli G (ed) (2002) Simulation des Kühlschmierstoffkreislaufs zur Optimierung einer umwelt- und ressourcenschonenden Produktionstechnik. DBU Forschungsbericht, FZ 13581. Shaker Verlag, Aachen

Priarone PC, Robiglio M, Settineri L, Tebaldo V (2016) Modelling of specific energy requirements in machining as a function of tool and lubricoolant usage. CIRP Ann 65:25–28. https://doi.org/10.1016/j.cirp.2016.04.108

Priarone PC, Robiglio M, Settineri L (2018) On the concurrent optimization of environmental and economic targets for machining. J Clean Prod 190:630–644. https://doi.org/10.1016/j.jclepro.2018.04.163

Pusavec F, Krajnik P, Kopac J (2010) Transitioning to sustainable production—Part I: Application on machining technologies. J Clean Prod 18:174–184. https://doi.org/10.1016/j.jclepro.2009.08.010

Rajemi MF, Mativenga PT, Aramcharoen A (2010) Sustainable machining: selection of optimum turning conditions based on minimum energy considerations. J Clean Prod 18:1059–1065. https://doi.org/10.1016/j.jclepro.2010.01.025

Rief M (2012) Vorhersagemodell für den Energiebedarf bei der spanenden Bearbeitung für eine energieeffiziente Prozessgestaltung. Dr.-Ing. dissertation, Universität Magdeburg. Shaker Verlag, Aachen

Robiglio M (2017) Modelling and optimization of machining processes towards economic and environmental sustainability. Dissertation, Politecnico di Torino

Schiefer E (2001) Ökologische Bilanzierung von Bauteilen für die Entwicklung umweltgerechter Produkte am Beispiel spanender Fertigungsverfahren. Dr.-Ing. dissertation, TU Darmstadt. Shaker Verlag, Aachen

Schlosser R (2013) Methodik zur Prognose der Nachhaltigkeit des Energie- und Stoffeinsatzes spanender Fertigungsprozesse. Dr.-Ing. dissertation, RWTH Aachen. Apprimus Verlag, Aachen

Schrems S (2014) Methode zur modellbasierten Integration des maschinenbezogenen Energiebedarfs in die Produktionsplanung. Dr.-Ing. dissertation, TU Darmstadt. Shaker Verlag, Aachen

Schudeleit T (2017) Total energy efficiency of factory-integrated machine tools. Dr. sc. dissertation, ETH Zurich, Zurich

Shokoohi Y, Khosrojerdi E, Rassolian Shiadhi B (2015) Machining and ecological effects of a new developed cutting fluid in combination with different cooling techniques on turning operation. J Clean Prod 94:330–339. https://doi.org/10.1016/j.jclepro.2015.01.055

Sihag N, Sangwan KS (2018) Development of a multi-criteria optimization model for minimizing carbon emissions and processing time during machining. In: 25th CIRP International conference on life cycle engineering, Copenhagen, Denmark, vol 69, pp 300–305. https://doi.org/10.1016/j.procir.2017.11.060

Soković M, Mijanović K (2001) Ecological aspects of the cutting fluids and its influence on quantifiable parameters of the cutting processes. J Mater Process Technol 109:181–189. https://doi.org/10.1016/S0924-0136(00)00794-9

Srinivasan M, Sheng P (1999) Feature-based process planning for environmentally conscious machining—Part 1: Microplanning. Robot Comput-Integr Manuf 15:257–270. https://doi.org/10.1016/S0736-5845(99)00017-4

Sutherland JW, Cao T, Daneil CM, Yue Y, Zheng Y, Sheng P, Bauer D, Srinivasan M, DeVor RE, Kapoor SG, Skerlos SJ (1997) CFEST: an internet-based cutting fluid evaluation software testbed. North Am Manuf Res Inst Soc Manuf Eng 25:243–248

Triebs JB (2015) Analyse, Optimierung und Simulation des Energieumsatzes dezentraler Fluidkreisläufe von Werkzeugmaschinen. Dr.-Ing. dissertation, RWTH Aachen. Apprimus Verlag, Aachen

Velchev S, Kolev I, Ivanov K, Gechevski S (2014) Empirical models for specific energy consumption and optimization of cutting parameters for minimizing energy consumption during turning. J Clean Prod 80:139–149. https://doi.org/10.1016/j.jclepro.2014.05.099

Wang Q, Liu F, Wang X (2014) Multi-objective optimization of machining parameters considering energy consumption. Int J Adv Manuf Technol 71:1133–1142. https://doi.org/10.1007/s00170-013-5547-z

Weber J, Weber J, Shabi L, Lohse H (2016) Energy, power and heat flow of the cooling and fluid systems in a cutting machine tool. In: 7th CIRP conference on high performance cutting, vol 46, pp 99–102. https://doi.org/10.1016/j.procir.2016.03.177

Winter M (2016) Eco-efficiency of grinding processes and systems. Dr.-Ing. dissertation, Technische Universität Braunschweig. Springer Verlag

Winter M, Thiede S, Herrmann C (2015) Influence of the cutting fluid on process energy demand and surface roughness in grinding—a technological, environmental and economic examination. Int J Adv Manuf Technol 77:2005–2017. https://doi.org/10.1007/s00170-014-6557-1

Xue H, Kumar V, Sutherland JW (2007) Material flows and environmental impacts of manufacturing systems via aggregated input–output models. J Clean Prod 15:1349–1358. https://doi.org/10.1016/j.jclepro.2006.07.007

Yan J, Li L (2013) Multi-objective optimization of milling parameters—the trade-offs between energy, production rate and cutting quality. J Clean Prod 52:462–471. https://doi.org/10.1016/j.jclepro.2013.02.030

Zein A (2012) Transition towards energy efficient machine tools. Dr.-Ing. dissertation, Technische Universität Braunschweig. Springer Verlag, Berlin

Zhang H, Haapala KR (2015) Integrating sustainable manufacturing assessment into decision making for a production work cell. J Clean Prod 105:52–63. https://doi.org/10.1016/j.jclepro.2014.01.038

Zhong Q, Tang R, Lv J, Jia S, Jin M (2016) Evaluation on models of calculating energy consumption in metal cutting processes: a case of external turning process. Int J Adv Manuf Technol 82:2087–2099. https://doi.org/10.1007/s00170-015-7477-4

Züst SD (2017) Model based optimization of internal heat sources in machine tools. Dr. sc. dissertation, ETH Zurich, Zurich

Chapter 4
Concept Development

This chapter focuses on the concept development, starting with the definition of requirements and presentation of the general concept structure (Sect. 4.1). The main part of this chapter is the modeling of input and output flows as well as the interactions within machining systems (Sect. 4.2). Based on these models technological, environmental and economic indicators are derived (Sect. 4.3) and integrated improvement measures are presented (Sect. 4.4). Finally, the presented concept is implemented within a prototypical software tool and the application cycle is presented (Sect. 4.5).

4.1 Framework for Concept Development

On the basis of the previous results and demonstrated research demand, the objectives and requirements are presented for this work. The proposed approach to meet them is illustrated by the concept structure in the following.

4.1.1 Definition of Objectives and Requirements

The previous chapter reviewed different research approaches focusing on the planning, assessing and improving of machining systems. These approaches differ regarding the addressed focus level and research object as well as used methods. Although a variety of different approaches have been investigated, a research gap regarding the comprehensive consideration of subsystems and components affected by the use of different cutting fluids strategies from a technological, environmental and economic perspective could be revealed. This research, therefore, aims at the development of a concept focusing on the interactions between process parameters, peripheral subsystems and cutting fluid strategies in order to support decision making

N. Madanchi, *Model Based Approach for Energy and Resource Efficient Machining Systems*, Sustainable Production, Life Cycle Engineering and Management,
https://doi.org/10.1007/978-3-030-87540-4_4

for existing as well as new machining systems without shifting of problems. For this purpose, the proposed concept has to follow specific requirements and is limited to certain system boundaries. Based on the evaluation criteria applied for the state of research, the specific requirements of this concept are outlined below.

Research scope

In view of the overall objective presented, the focus is on the development of a model based concept to plan and assess energy and resource efficient machining systems. The general scope of the concept is determined by the following requirements:

- Creating a transferable concept that can be used for various configurations of machining systems to ensure a broad application.
- Consideration of possibilities to extend the concept by further subsystems and components to areas such as the application of other cutting fluid strategies that are not yet taken into account.
- Development of a modular structure that enables the use or elimination of certain elements.
- Implementation of the concept within an intuitive software tool using standard software.

Research object

Certain subsystems and components are especially critical with regard to the objectives of the concept. It is initially important to define the system boundaries and to limit the concept to the following aspects:

- Integrated consideration of machining system elements covering all subsystems affected by the use of cutting fluids.
- Examination of critical components regarding e.g. energy and resource consumption at a higher level of detail.
- Consideration of general turning, milling and grinding processes as individual machining processes or within a process chain. As not the entire machine tool is affected by cutting fluid strategies, the focus is on the machining process itself.
- Incorporation of cutting fluid strategies using alternative types of fluids focusing on water-based and oil-based fluids as well as alternative amounts of fluids focusing on flooding and dry machining.

Research method

The development of such a model-based planning and assessment concept requires the use of various methods and tools. These need to fulfill the following requirements:

- Modeling of all relevant energy and resource flows between the considered elements of the machining system.
- Implementation of generally applicable and transferable models that can be used with little effort. Preference of using semi-empirical models, which are based

on physical laws as well as empirical coefficients that are widely available in literature.

- Validation of the model quality on the basis of experimental data, aiming at a sufficient level of average accuracy between calculated and measured results of around 5–30% (see Sect. 2.5).
- Characterization of workpiece and process quality for an evaluation of technology related performance indicators.
- Consideration of energy and resource flows within the machining system for an evaluation of environment related performance indicators.
- Linking energy and resource flows as well as other system characteristics to costs for an evaluation of economic related performance indicators.
- Consideration of technical and organizational improvement measures.

The proposed concept should be applicable by different potential stakeholders and users for a variety of purposes. At production planning level the concept is intended to support the selection of process parameters as well as the regulation of the peripheral subsystems by e.g. the exhaust air flow or cutting fluid volume flow. This applies to both existing and new machining systems. It could further be used to choose a cutting fluid strategy and to decide on centralized or decentralized peripheral subsystems within the greenfield planning of new machining process chains. For a supplier of machine tools and peripheral systems, the insights into the interactions could also help the dimensioning of different elements such as the piping and pumps. Further, a cutting fluid supplier and developer could use the concept to support the development of new fluid formulations but also to demonstrate the impact of their product in comparison to alternatives to their customers. These demonstrate only exemplary applications to which the concept is not limited and it can also cover further applications.

4.1.2 Concept Structure

The proposed approach to fulfill the presented requirements and objectives is composed of four modules as illustrated in Fig. 4.1. The first module provides the basis of the concept with the *System description*. For this purpose, the machining system is divided into relevant subsystems and components. Within the concept, cutting fluid connects these elements of the machining system and results in interactions between them. For each system element, the functions and relevance are specified. Based on the description of the elements, the second module aims at the *Modeling of machining system elements* with the preferred use of semi-empirical models. The models allow to determine the energy and resource flows or technological results from the elements separately. This modular approach enables the combination of the elements into different machining systems and to evaluate the impact of different cutting fluid strategies, process parameters and configuration

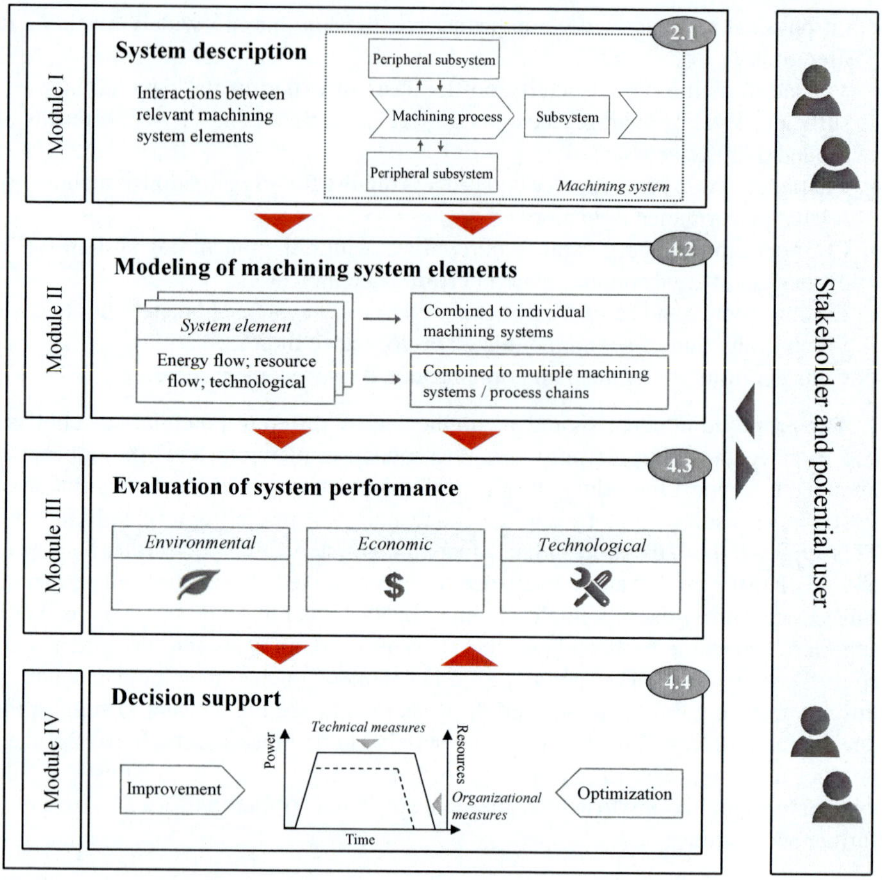

Fig. 4.1 Concept procedure for energy and resource efficient machining systems

changes on them. In order to ensure model quality and a sufficient level of accuracy, this module also aims at the analysis of experimental data to validate and adapt some of the models. The *Evaluation of system performance* is implemented in the second module and relates to the overall machining system consisting of the different elements. Their modeled energy and resource flows are combined to assess the impact from an environmental and economic perspective, while the technological perspective is assessed by modeled product- and process-specific indicators. Through the model-based approach, changed configurations of single elements as well as strategy changes at system level are integrated and can be directly evaluated. This integration is the basis for the *Decision support,* which is targeted in the fourth module. A set of technical improvement measures with a focus on hardware as well as organizational measures with a focus on process planning are provided. The measures are intended to decrease environmental and economic impact without endangering the technological results or improving them. Overall, the concept strives to meet the requirements of the stakeholders and potential users and to create the required transparency in order to achieve added value for them. For a simplified application, the concept is implemented in a tool and the general application cycle for the user is demonstrated.

The modules of the concept for a model-based approach towards energy and resource efficient machining systems are explained in the following sections, whereas the system description is generally based on the results from the second chapter.

4.2 Modeling of Machining System Elements

In view of the development of a model-based approach, the modeling of the machining system elements is a central aspect of the concept and defined as the second module of the concept procedure. The general interactions between the relevant subsystems are already presented in Sect. 2.1. With regard to the described requirements, it is also necessary to model their energy and resource flows as visualized in Fig. 4.2.

Based on this system description, the exhaust air system is defined by the energy demand to remove air from the machine tool and to filter it from cutting fluid emissions. Due to its importance in terms of occupational health and safety, it is assumed that exhaust air systems are always required independent of the cutting fluid strategy but they can be designed as a centralized or decentralized system. The cutting fluid system is mainly described by the use of cutting fluid and energy demand to supply and maintain it. The system can also be designed as a centralized or decentralized system, but in this case, a cutting fluid system is not required for each strategy such as dry machining. Likewise, a de-oiling of chips is only required when cutting fluids are used. As other processes regarding the scrap processing are not affected by the cutting fluid strategy, only the de-oiling process and its consumption of energy and dragged-out cutting fluids are considered. The tool preparation focuses on the use of energy and materials to produce and prepare the tools. In this case, the wear is critical as it determines the number of required tools or preparation cycles for the machining of a specific number of workpieces. Subsequent to the machining process, a cleaning process might be required depending on the cutting fluid strategy or following process. The cleaning process is defined by the use of energy and cleaning fluids. With regards to the machining system, another central element is the machining process itself. In this case, the cutting energy is taken into account as a relevant energy flow, because it depends on the cutting fluid strategy and parameter setting.

In addition to these direct energy and resource flows, the modeling of further aspects such as material removal rate, processing time as well as the surface roughness is also required for the subsequent evaluation of the system.

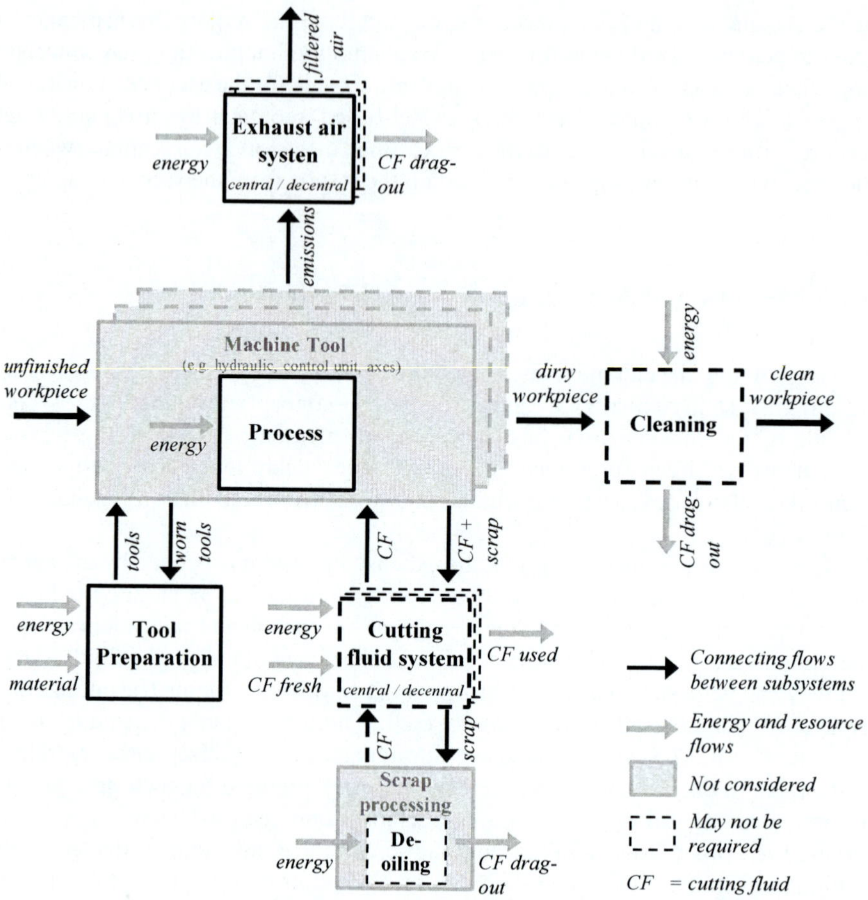

Fig. 4.2 Energy and resource flows within a machining system (Madanchi et al. 2019b)

4.2.1 Description of Procedure

In order to model the technological correlations as well as the energy and resource flows within the considered machining system, the use of well-established models from literature is preferred. In case these are not available, own empirical approaches are used. The specific approach is presented in the following section. To derive the required empirical models on the one hand but also to verify the applicability of some existing models on the other hand, experimental investigations are carried out. The experimental setup and approach is further presented in the following. The results are analyzed using the coefficient of determination (R^2), which indicates the agreement between the model and observed data. The R^2-value ranges between 0 and 1, where 0 means that none of the data distribution around its mean value are determined by the model and 1 means that the entire data distribution is determined. The closer the value to 1, the higher the accuracy of the model. The R^2-value is determined by Eq. 4.1 based on the modeled ($\widehat{Y}_i$), observed (Y_i) and mean observed values ($\overline{Y}_i$) (Kleppmann 2011):

$$R^2 = \frac{\sum_{i=1}^{n} (\widehat{Y}_i - \overline{Y}_i)^2}{\sum_{i=1}^{n} (Y_i - \overline{Y}_i)^2} \tag{4.1}$$

Further, the relative error (E) is also considered, which indicates the mean value of relative deviation between modeled and observed results. It is determined according to the following equation:

$$E = \frac{1}{n} \sum_{i=1}^{n} \frac{\left| Y_i - \widehat{Y}_i \right|}{Y_i} \tag{4.2}$$

The E-value also ranges between 0 and 1, but here a lower value is targeted as it indicates a lower relative error.

4.2.1.1 Modeling Approach

The described elements of machining systems are to be transferred into simplified models. These models are a mathematical representation of reality and can be used to conduct a simulation of experiments. According to the requirements, it is the aim to use transferable models, which can be applied to a variety of different process and system designs. It should further be possible to use these models with little effort. In accordance with the requirements, it is therefore preferred to use semi-empirical models based on physical laws combined with empirical coefficients. In view of the concept's objective, it is required that the models take the impact of cutting fluids and their interactions into account.

With regard to the model development for the different subsystems and related components, existing modeling approaches have been collected and investigated. Figure 4.3 visualizes an overview of approaches from literature that are considered for the concept. These models present the basis but are usually adapted and extended by other models and additional coefficients.

Accordingly, the used models are briefly described and summarized below, whereas the detailed description of the models is presented for each subsystem in the following sections.

The cutting force and power is determined based on the cutting force model of Kienzle, which is a well-established model (Kienzle 1952). It is based on coefficients to determine the specific force to remove a specific chip size. Due to the wide use, the coefficients for various workpiece materials and tools are already available in literature which enables a flexible and simple model use. Over the years, the model

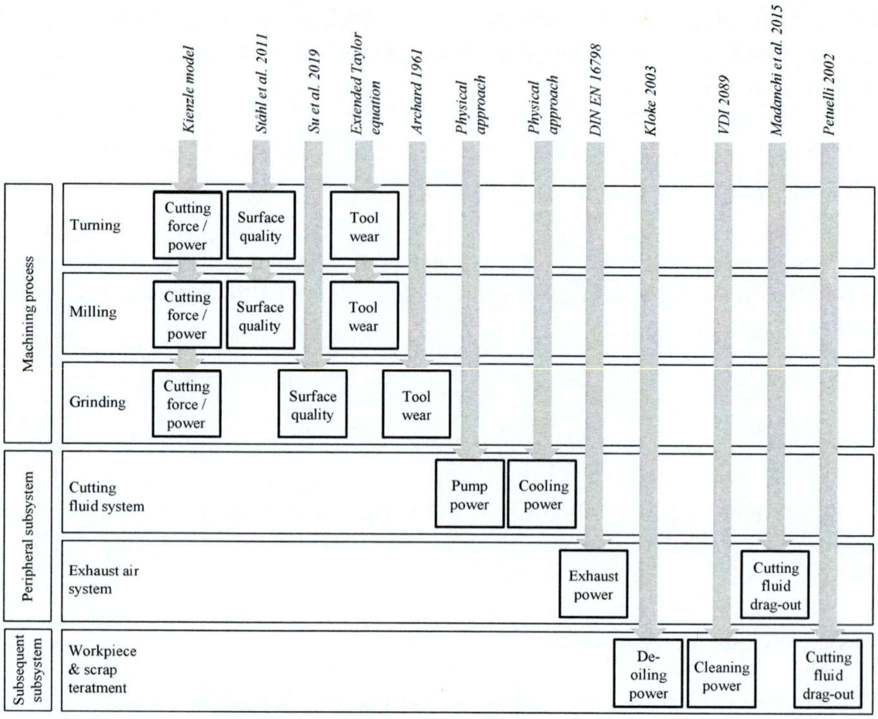

Fig. 4.3 Overview of implemented modeling approaches

has been extended by additional correction factors that also consider the cutting fluid impact. Further, this approach is applicable to turning, milling as well as grinding.

The workpiece quality is represented by the arithmetic mean roughness, which is determined based on Ståhl et al. (2011) for tuning and milling processes. It considers the feed rate as well as the nose radius of the cutting tool and is further extended by a correction factor for cutting fluids. The determination of the surface roughness for grinding processes differs and is based on Su et al. (2019) considering the undeformed chip thickness.

The tool wear of turning and milling processes is modeled by the extended Taylor equation. For this approach, there are also constants available in literature for different materials and tools. The approach for grinding is again different. In this case, the wear volume of the grinding wheel is determined using an approach by Winter (2016) based on Archard (1961). The cutting fluid impact is taken into account by the friction coefficient.

Regarding the cutting fluid system, the focus is on the determination of the pump power, which is used to supply, circulate and filter the cutting fluid. This is based on a general physical model, but includes other models for example to determine the hydraulic and motor efficiency as well as the pressure loss due to the piping. Further, the cooling power is modeled to maintain the cutting fluid at a constant temperature. For this approach the heat input is modeled due to the pumps based on physical relations used by Züst et al. (2013) and due to the cutting process based on simplified assumptions according to Knobloch (1979). Besides heat input, heat losses due to heat radiation and convection are also modeled based on physical relations as presented by VDI 2055 (2019).

For the exhaust air system the electric fan power is particularly relevant and can be modeled based on DIN EN 16798-3 (2017). The required exhaust air flow depends on the machining process and cutting fluid strategy and is modeled according to VDI 3802-2 (2012). Further, this subsystem is also connected with the drag-out of cutting fluid emissions which are determined based on the empirical approach of Madanchi et al. (2015a).

The workpiece and scrap treatment is characterized by the power demand to de-oil the scrap material and clean the workpieces. For the de-oiling process, the power demand of a vertically rotating centrifuge is modeled based on the physical approach according to Kloke (2003). The cleaning process is modeled according to VDI 2089 (2010), an approach that determines the heating of swimming pools. Additionally, the drag-out of cutting fluid attached to the scrap and workpiece surface is relevant. This is modeled based on to the empirical approach presented by Petuelli (2002).

4.2.1.2 Experimental Approach

For the purpose of data collection and experimental validation of model quality, an exemplary workpiece with an initial diameter of 85 mm and a length of 85 mm is machined. The production of the workpiece combines turning, milling and grinding operations within a process chain (see Fig. 4.4). Workpieces made of steel with the designation 1.3505 (100Cr6) and 1.1191 (C45) are used and hardened after the milling process in preparation for grinding. The hardening process is not within the scope of this research.

Regarding the experimental design, different methods are considered. Especially with the focus on model validation, the method one-factor-at-a-time is used, meaning that one parameter is changed while all other parameters remained the same. However, with the aim to derive an empirical model, statistically designed experiments are performed. In this case, a central composite design is considered, in order to reduce the number of experiments and to increase the statistical significance. The used experimental setups for the machining processes are presented in the following.

Fig. 4.4 Exemplary process chain for the experimental verification

Machining system - Turning

The turning operation is conducted on a Spinner TC600 CNC universal turning lathe. For this machine tool the maximum spindle power is 34 kW with a maximum spindle speed of up to 5000 min^{-1}. The used tool is a universal insert CCMT 09T304 suitable for wet and dry machining. The machine tool is connected to a cutting fluid system with a filling volume of around 180 l and a pump pressure of up to 9.8 bar is used to supply the cutting fluid. Further, an exhaust air system with two operation settings of 600 and 1000 m^3/h is also powered and controlled by the machine tool. For the experiments, a mineral oil-based grinding oil with a kinematic viscosity of 12.0 mm^2/s and a mineral oil-based emulsion with a concentration of 7.5% and a kinematic viscosity of around 1.0 mm^2/s are used. Besides the use of cutting fluids, dry machining is also investigated.

For simplification, only selected results for the external turning process are further presented.

Machining system - Milling

A DMG Mori DMU 100 monoBlock is used to perform the milling operation. This 5-axis machining center has a maximum spindle power of 15 kW and a maximum spindle speed of up to 12.000 min^{-1}. Basically three different milling operations are performed within the process chain and therefore three different solid carbide tools with a diameter of 10, 8 and 3 mm are used. For the flooding strategy a cutting fluid system with a filling volume of around 200 l and a pressure of 4.8 bar is used to supply the mineral oil-based emulsion with a concentration of 7.5%. Similar to turning, dry machining is also examined as part of this experimental investigation. The machine tool also powers and controls an exhaust air system with an exhaust air flow of 800 m^3/h.

For simplification, the focus of the following results and evaluations is particularly on the 10 mm slot milling process.

Machining system - Grinding

The grinding operation is performed on a Studer S120 internal cylindrical grinding machine with a maximum spindle power of 10 kW and a maximum spindle speed of up to 60,000 min^{-1}. For the experiments, a cubic boron nitride (cBN) grinding wheel with a width of 15 mm and a diameter of 40 mm is used (specification B126 M8 VD 49VV 80). A cutting fluid system with a filling of around 100 l and a pressure of 6.7 bar is used to supply the cutting fluid. For grinding, the same mineral oil-based grinding oil and mineral oil-based emulsion with a concentration of 7.5% are investigated. The machine tool is further connected to an exhaust air system with an exhaust air flow of 2500 m^3/h.

Measurement devices and data analysis

Different measurement devices are used to analyze the process performance. A three-phase power meter type PPC-3 from Load Controls® is used to continuously measure the power consumption of the respective machining system. The process forces are measured using two 3-component force sensors type 9251A from Kistler® for grinding and a measuring platform type 5011 from Kistler® for milling. This measurement method uses piezoelectric sensors and increases their electric charge by an amplifier type 5747A from Kistler®. The signals from the force and power metering are recorded and processed with a system based on LabView® from National Instruments. The collected data, which are recorded at high resolution, are then analyzed using the flexpro® software from Weisang.

Regarding the workpiece quality, the surface roughness is measured after each machining process at five different positions on the workpiece. For this purpose the tactile surface measurement device T1000 Hommel-Etamic® from Jenoptik is used to detect the the arithmetic mean roughness value.

The particle measurement during the experiments is performed using the Handheld 3016® from Lighthouse. It measures particles from a size of 0.3 up to 25 μm and classifies them into six groups (>0.3 μm; >0.5 μm; > 1.0 μm; > 5.0 μm; >0.0 μm; >25.0 μm).

The regression coefficients for empirical models are determined based on the experimental data. For the design of these experiments and determination of regression coefficients, the software Minitab® is used.

4.2.2 Machining Process Model

In view of the concept structure, the third module focuses on the modeling of the different machining system elements. For the machining process, the electric power demand, surface roughness and tool wear are modeled for turning, milling and grinding processes. For this purpose, different models from literature are used (see Fig. 4.3) and combined with additional models and coefficients.

4.2.2.1 Power Demand

The power demand of the machine tool and the cutting process are particularly relevant. For the analysis, different machine tool states have to be distinguished. At the ramp-up state the machine tool is started up after being switched-off and is then operational in the standby state. The processing state described the state of actual material removal. In order to visually analyze the electric power and energy demand of individual components in relation to the operational state, an energy breakdown analysis of the respective machine tool system is initially performed on the basis of the experiments. The schematic structure of energy breakdowns is illustrated in Fig. 4.5. It displays how components are assigned to the load profile and the differentiation between the operational states of the machine tool. This visualization enables the identification of hotspots as well as the first derivation of improvement potentials (Zein 2012).

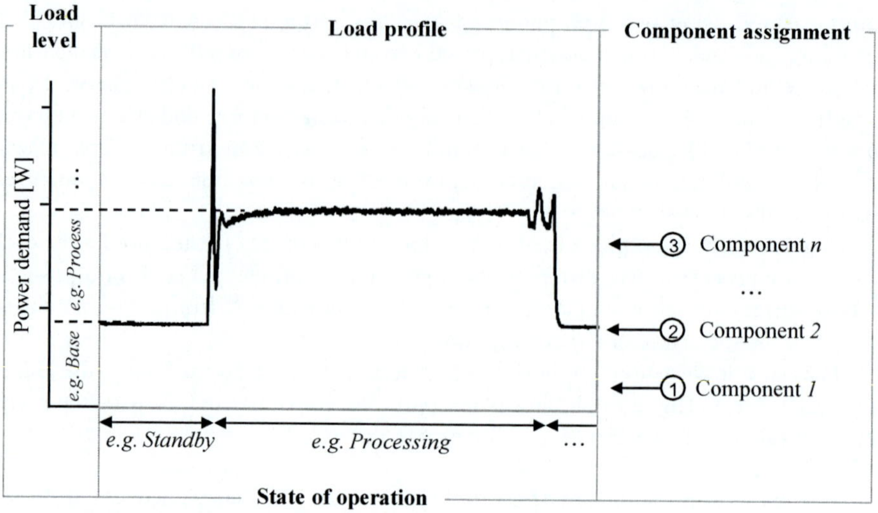

Fig. 4.5 Schematic structure of an energy breakdown analysis

Turning process

The energy breakdown analysis of the turning machining system is presented in Fig. 4.6 and indicates the allocation of different machine tool components to the energy demand. It also demonstrates the machining profile performing individual cuts with a depth of 0.5 mm and resulting in a total depth of 3.5 mm.

During the ramp-up state, the components of the machining system are activated sequentially, starting with the control unit and followed by further components such

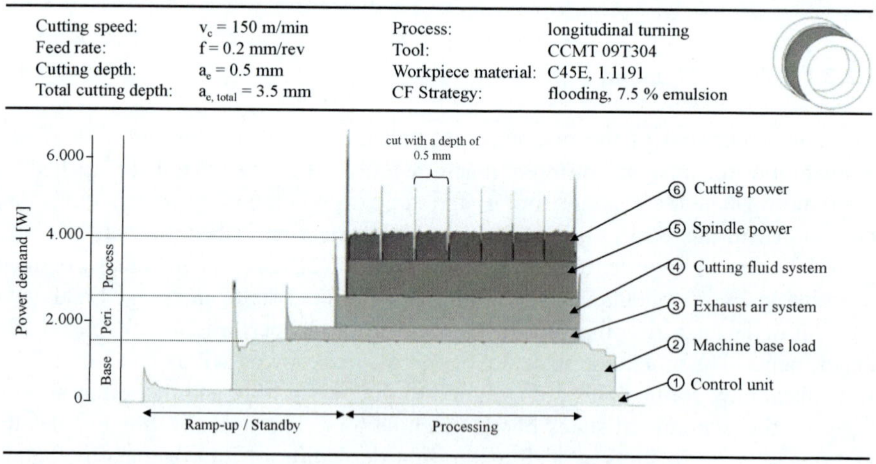

Fig. 4.6 Energy breakdown analysis of the turning machining system

as the hydraulic unit. After ramp-up the machine tool is operational and in standby mode. At this state the machine tool operates at base load, which accounts for about one third of the total load and is usually constant. The peripheral systems of exhaust air and cutting fluid system account for almost another third. Due to the long ramp-up phase of the exhaust air system it remains active, while the cutting fluid system is only used during the processing state. The process power consisting of the spindle and material removal accounts for the remaining third of the total load. However, in contrast to the base load, the actual energy involved in cutting is not constant and strongly depends on aspects such as process parameters, workpiece material or cutting fluid. Its share of the total machine tool load varies and can generally also account for 10–70% (Dahmus and Gutowski 2004). The cutting power is thus critical for the concept and modeled in the following.

As a model basis the interaction between the cutting parameters as well as the basic kinematics of a turning process are demonstrated in Fig. 4.7.

Figure 4.7 demonstrates the impact of process parameters on the chip formation and resulting cutting force. Depending on the working cutting edge angle (κ), the feed rate (f) as well as the depth of cut (a_e) influence the surface area of the undeformed chip. Overall, the cutting power (P_c) is modeled based on the cutting force (F_c) and the cutting speed (v_c):

$$P_c = F_c \cdot v_c \tag{4.3}$$

The cutting force is usually determined based on the cutting force model of Kienzle (1952).

$$F_c = k_{c1.1} \cdot \mathrm{b} \cdot h_{cu}^{(1-mc)} \tag{4.4}$$

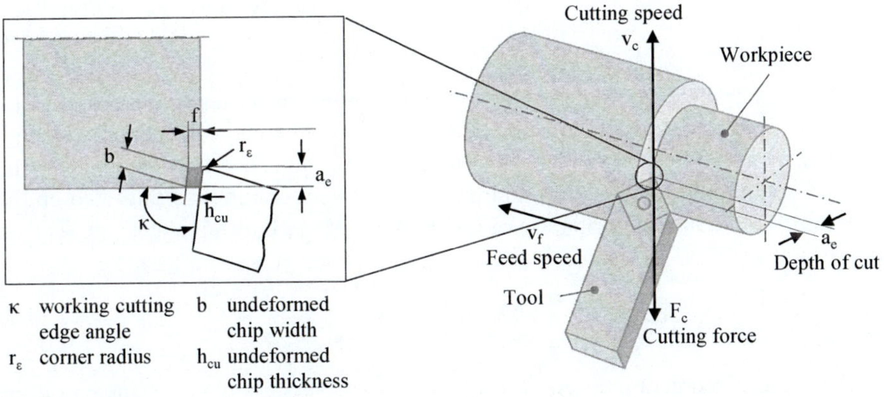

Fig. 4.7 Geometric and kinematic relationships of external turning [based on Denkena and Tönshoff (2011)]

This is an empirical approach, which describes the specific force ($k_{c1.1}$) that is required to detach a chip with the cross section of 1 mm^2 ($h = b = 1$ mm). The equation was approximated by a straight line, where the exponent $(1 - mc)$ indicates the gradient. Values for the specific cutting force and exponent have been empirically determined by turning experiments for common workpiece materials and can therefore be taken from tables provided in different literature (König and Essel 1982; Victor 1969). Following the trigonometric relationship presented in Fig. 4.7, the undeformed chip width (b) and chip thickness (h_{cu}) can be calculated based on the following equations:

$$b = \frac{a_e}{\sin\kappa} \tag{4.5}$$

$$h_{cu} = f \cdot \sin\kappa \tag{4.6}$$

However, the calculation of the cutting force is rather simplified and depends on further parameters in practice. The equation has, therefore, been extended by additional correction factors in the past (see Eq. 4.7) (Schönherr 2002). This extended model is widely accepted and applied in different research approaches such as by Eisele (2014).

$$F_c = k_{c1.1} \cdot b \cdot h_{cu}^{(1-mc)} \cdot K_\gamma \cdot K_m \cdot K_{vc} \cdot K_{vk} \cdot K_{tw} \cdot K_{cf} \tag{4.7}$$

Specific values for the correction factors or equations to determine them are proposed in literature by Schönherr (2002). Accordingly, the correction factor for the rake angle (K_γ) is determined by the following equation with the possible reference values (γ_n) of 6° for steel and 2° for casting materials:

$$K_\gamma = 1 - \frac{\gamma_n - \gamma_0}{100} \tag{4.8}$$

Different friction coefficients for different cutting materials are considered by the correction factor (K_m). It is for example 1 for cemented carbide, 1.2 for high speed steel and 0.9 for ceramics. Moreover, the cutting speed is considered by the correction factor (K_{vc}), as it has an impact on the temperature and thus also on the tool due to a possible deformation of the cutting edge or built-up edge.

$$K_{vc} = \left[\frac{100}{v_c}\right]^{0.1} \tag{4.9}$$

The workpiece geometry and type of machining process have an impact on the chip formation and therefore also on the cutting force. The impact is considered by the correction factor (K_{vk}), which can be 1 for external turning and 1.05 + 1/d for internal turning and milling. Tool wear during processing is considered by the correction factor (K_{tw}) The friction between tool and workpiece further depends on the used cutting fluid strategy and its cooling and lubricating effect. The correction factor (K_{cf}) is 1 for dry machining, 0.94 for water-based emulsions and 0.8 for the use of oil-based cutting fluids.

As described in the procedure this model is applied to a case study and the cutting power is determined for different process parameters. In order to validate the model quality, these values are compared with the experimentally measured values. The results are presented in Fig. 4.8.

The results demonstrate an increasing cutting power with increasing process parameters. By increasing feed rate and cutting depth, the chip size increases as well and requires higher forces to be detached from the workpiece. The cutting depth demonstrates the highest impact within the considered range of parameters. Overall, the experimental results are only slightly higher for dry machining than for wet machining in the case of few parameter settings. Otherwise they are rather similar. The modeled results, by contrast, differ slightly in relation to the correction factor for the cutting fluid strategy. It is generally assumed that the cutting fluid in the contact zone leads to a reduced contact length between tool and workpiece and friction coefficient, which consequently reduces the cutting force (Sanchez et al. 2013). Alternatively, it is also possible that cooling may reduce the temperature of the chip and hence increases its strength, which results in increased cutting forces. The influence of process parameters and especially the cutting speed is therefore decisive (Klocke 2011). Nevertheless, the high R^2- and E-values generally indicate a good and sufficient consistency between the modeled and experimentally measured results.

Milling process

According to the exemplary process chain, the turning process is followed by a milling process. The energy demand of the milling machine tool and its peripheral subsystems is again visualized by an energy breakdown analysis (see Fig. 4.9). In

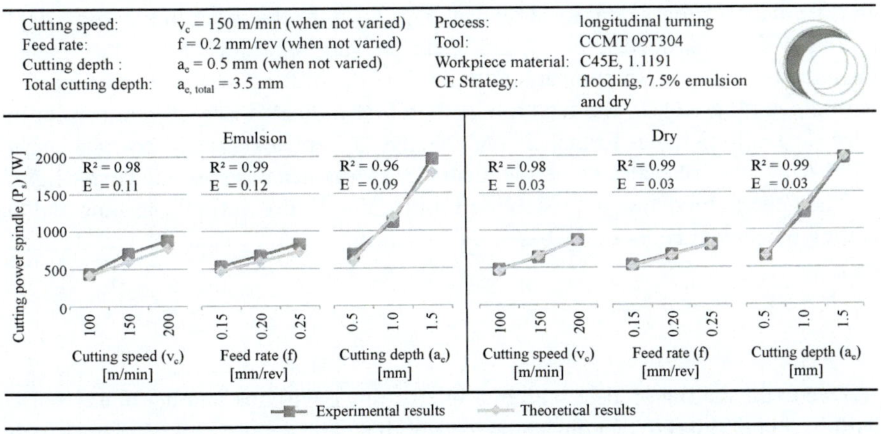

Fig. 4.8 Comparison of modeled and experimental cutting power of turning processes

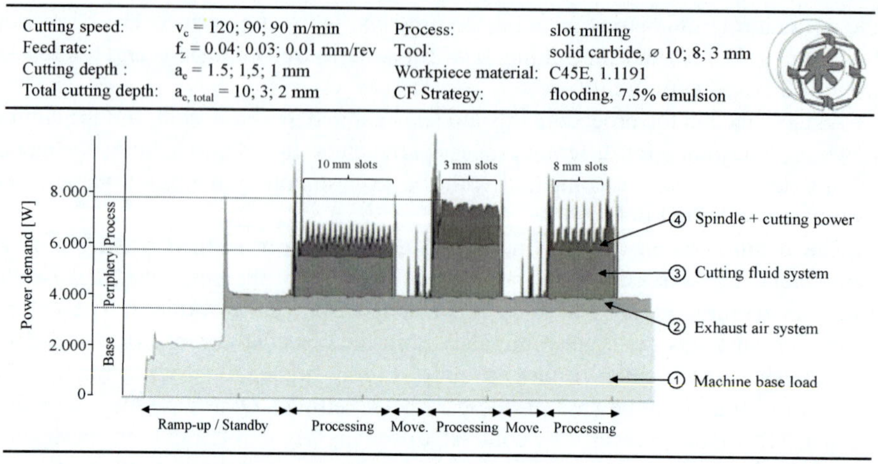

Fig. 4.9 Energy breakdown analysis of the milling machining system

this case, the analysis involves three different milling operations with three different milling tools. The tool diameter has a significant effect on the spindle speed and power even without cutting. Hence, the 3 mm milling tool requires a higher spindle speed and power compared to the 8 and 10 mm tools. The base load of the machine tool, in contrast, is again rather independent from process and tool parameters and accounts for about 30–50% of the total load. Regarding the peripheral subsystems, the exhaust air system is again operated independently from the machining process and is continuously active. The cutting fluid, by contrast, is only supplied to the machine tool when the spindle is operated.

With regards to the machine tool and process perspective, it is again particularly relevant to model the actual cutting power as it depends on the selected process parameters. It can be modeled based on the cutting force, which depends on the chip formation process demonstrated in Fig. 4.10. In this case, the cutting depth (a_e) as well as the feed rate (f_z) are again relevant regarding the surface area of the undeformed chip. In comparison to turning, the milling tool is characterized by multiple cutting edges. The feed rate is therefore indicated as feed rate per cutting edge. The milling tool is further characterized by a diameter (D) and a certain cutting width (a_p), which can also be smaller than the tool diameter as presented in the figure.

The general modeling approach used for turning is also valid for milling and the cutting power can be modeled based on the cutting force (F_c) and the cutting speed (v_c):

$$P_c = F_c \cdot v_c \cdot z_c \tag{4.10}$$

Due to the cutting process with multiple cutting edges, the number of edges used within a cut (z_c) has to be considered as well. It can be determined in relation to the total number of cutting edges (Z) and the cutting direction angle (φ_c) demonstrated in Fig. 4.10:

$$z_c = Z \cdot \frac{\varphi_c}{2\pi} \tag{4.11}$$

The cutting direction angle further depends on the ratio of cutting width to tool diameter and can be determined as follows:

$$\varphi_c = \begin{cases} \arccos\left(1 - \frac{a_p}{\frac{D}{2}}\right), a_e/D < 0.5 \\ \pi - \arccos\left(\frac{a_p}{\frac{D}{2}} - 1\right), a_e/D \geq 0.5 \end{cases}$$

In order to determine the cutting force, the Kienzle model as well as the cutting coefficients can be applied again, taking the individual geometric characteristics into account (Victor 1969). Moreover, the extended correction factors can also be applied and determined as described for the turning process (Schönherr 2002):

$$F_c = k_{c1.1} \cdot b \cdot h_{cu}^{(1-mc)} \cdot K_\gamma \cdot K_m \cdot K_{vc} \cdot K_{vk} \cdot K_{tw} \cdot K_{cf} \tag{4.12}$$

The undeformed chip width is determined using the trigonomic relationship between the depth of cut and the angle of the working cutting edge (see Fig. 4.10):

$$b = \frac{a_e}{\sin\kappa} \tag{4.13}$$

Although the average undeformed chip thickness (h_{cu}) also depends on the trigonomic relationship, it has to be extended compared to turning due to the multiple cutting edges (Schönherr 2002):

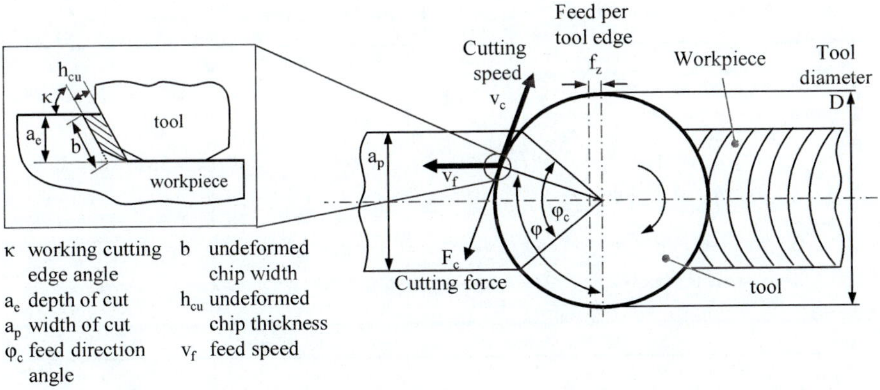

Fig. 4.10 Geometric and kinematic relationships of face milling [based on Denkena and Tönshoff (2011)]

$$h_{cu} = \frac{360^\circ}{\pi \cdot \varphi_c} \cdot f_z \cdot \sin\kappa \cdot \frac{a_p}{D} \tag{4.14}$$

This approach for calculating the cutting power of milling processes is already common in literature and for example used by Rief (2012).

The presented modeling approach was again applied to the presented workpiece and compared with the experimentally measured cutting power. The results for milling the 10 mm slots of the workpiece are presented in Fig. 4.11.

The impact of varying cutting speed, feed rate and cutting depth are analyzed for wet machining using a mineral oil based emulsion and dry machining. In all cases, the results demonstrate an increasing cutting power with increasing parameter settings. Similar to the turning results in Fig. 4.8, the cutting depth shows the highest impact on the results for the examined parameter variation. An increasing feed rate and cutting depth lead to an increasing undeformed chip surface area that has to be removed within the milling process and therefore requires a higher cutting force and power. The experimental and theoretical results demonstrate an increasing cutting power with increasing cutting speed, although the calculated results are slightly higher than the measured ones. In this case, it could be relevant that an increasing cutting speed leads to higher temperatures in the contact zone, which enables the material to be removed more easily and consequently reduces the cutting force. This effect is only indirectly considered within the models by the correction factor K_{vc} (Schönherr 2002). The impact of the cutting fluid strategy on the chip formation is similar to the turning process and the results for it are rather similar. Overall, the comparison of the results demonstrates a good level of accuracy with high R^2-values and a sufficiently low relative error.

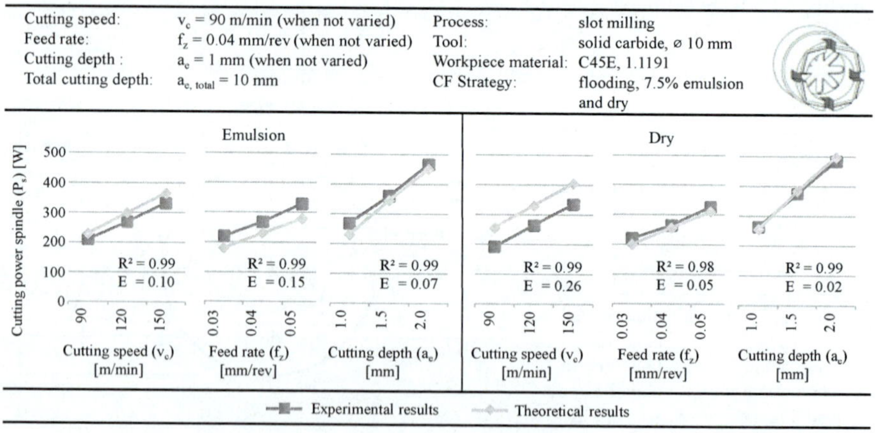

Fig. 4.11 Comparison of modeled and experimental cutting power of milling processes

Grinding process

The milling process is followed by a grinding process according to the process chain. For this case study, a cylindrical grinding process of the internal surface is performed. The energy demand of the used machine tool and system is again visualized by an energy breakdown analysis, which is shown in Fig. 4.12.

For the grinding system, the ramp-up phase starts again by activating the control unit and other components of the machine base load such as hydraulic and lubrication unit. Similar to the turning system, it accounts for about one third of the total load. The peripheral subsystems include the exhaust air system and cutting fluid system, which are activated together with the spindle. Overall, the ratio of cutting power for the actual material removal to total power is again comparatively very low.

The cutting power strongly depends on the chip formation process, which is presented in Fig. 4.13. It illustrates a process in which a grinding wheel with the diameter (d_g) rotates and removes a chip with a maximal surface area ($A_{cu,\max}$) from a workpiece with the speed (v_w). The contact area between grinding wheel and workpiece is characterized by the arc length (l_g). In comparison to the milling process, which already uses several cutting edges, grinding is characterized by an even higher number of cutting edges. Thus, each grain has to be considered and its resulting maximal undeformed chip width ($b_{\max}$) and thickness ($h_{cu,\max}$).

The determination of the cutting power is therefore similar to the milling process and can be modeled based on the cutting force (F_{czm}) and the cutting speed (v_c), but further depends on the number of grains (z_c) used for the grinding process (Schönherr 2002):

$$P_c = F_{czm} \cdot v_c \cdot z_c \tag{4.15}$$

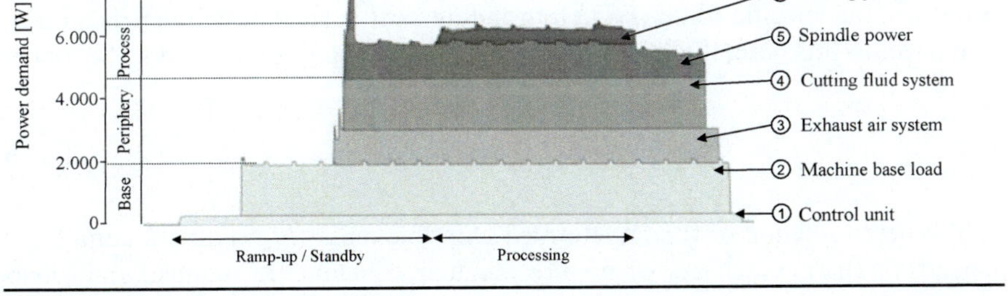

Fig. 4.12 Energy breakdown analysis of the grinding machining system

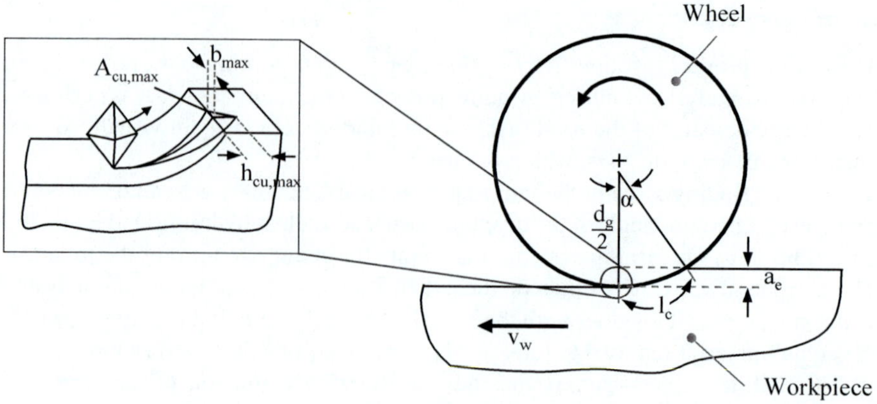

Fig. 4.13 Geometric and kinematic relationships of grinding [based on Guo and Malkin (2000), Winter (2016)]

Accordingly, the number of grains in contact have to be determined. This number highly depends on the characteristics of the grinding and conditioning process (Torrance and Badger 2000). In literature, various approaches are proposed to determine the number of grains. Winter (2016), for example, considered the approaches by Yegenoglu (1986) and Zitt (1999) as well as the approach by Torrance and Badger (2000) to also consider the conditioning process. It depends on the particular geometry of the grain and the lattice structure. This concept, however, aims at a generally valid approach that focuses on the relevant influencing factors of the grinding process. These are the cutting depth as well as the cutting and workpiece speed (Winter 2016). Thus, the following model is used which is also proposed in the basic literature (Malkin and Guo 2008; Schönherr 2002):

$$z_c = \frac{d_g}{\lambda_{ke}} \cdot \sqrt{\frac{a_e}{d_g \cdot (1 \pm \frac{d_g}{d_w})}} \tag{4.16}$$

The cutting force per grain (F_{czm}) can then be determined using the Kienzle model in accordance to turning and milling, but using the empirical correction factor (K_{cv}) to transfer the specific force ($k_{c1.1}$) to grinding as it is smaller and depends on the grain size. In this case, the width (b) corresponds to the grinding width (Schönherr 2002):

$$F_{czm} = k_{c1.1} \cdot b \cdot h_{cu}^{1-m} \cdot K_{cv} \cdot K_{cf} \tag{4.17}$$

The determination of the undeformed chip thickness (h_{cu}) is more complex. It depends on the grain shape, where a rectangular, triangular or rounded grain shape leads to different chip sections and can be determined differently (Rowe 2009). In contrast to the detailed approaches, there are also more general approaches, which provide an empirical reference value for the chip thickness depending on the process (Meister 2011). For this concept, however, the chip thickness is determined based on the following equation, which considers the grains in general, but takes into account the impact of process parameters and characteristics. The effective grain spacing (λ_{ke}) is taken from empirical reference values from literature (Schönherr 2002):

$$h_{cu} = \begin{cases} \frac{\lambda_{ke}}{q} \cdot \sqrt{a_e \cdot \left(\frac{1}{d_g} + \frac{1}{d_w}\right)}, & external\ grinding \\ \frac{\lambda_{ke}}{q} \cdot \sqrt{a_e \cdot \left(\frac{1}{d_g} - \frac{1}{d_w}\right)}, & internal\ grinding \\ \frac{\lambda_{ke}}{q} \cdot \sqrt{\frac{a_e}{d_g}}, & surface\ grinding \end{cases} \tag{4.18}$$

In contrast to turning and milling, the workpiece as well as the tool are usually in motion during grinding. The speed ratio (q) between grinding wheel and workpiece is therefore determined as followed:

$$q = \frac{v_s}{v_w} \tag{4.19}$$

The application of this approach to the presented case study leads to the theoretical cutting power presented in Fig. 4.14. It is again compared to the experimentally determined cutting power.

The results again demonstrate an increasing cutting power with increasing process parameters. All of the considered process parameters have an impact on the undeformed chip thickness. While the trend is still similar and shows rather high R^2-values, the relative error between the results is higher. For the results using emulsion, the E-values are still below 30%, but the results using oil are on average above it. This

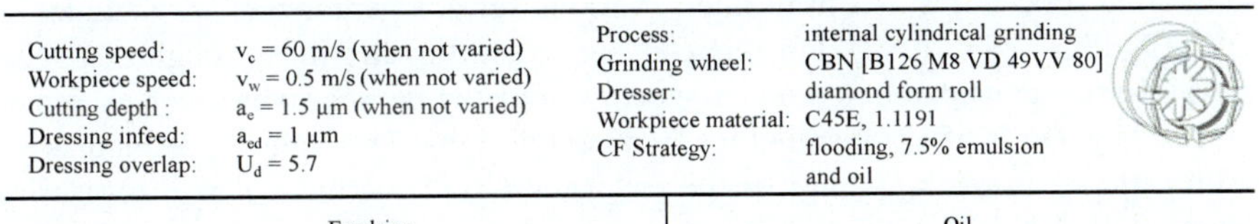

Fig. 4.14 Comparison of modeled and experimental cutting power of grinding processes

might be due to the additional empirical correction factor that is required to transfer the specific force to the grinding process. Moreover, the generic models exclude some influencing factors such as grain shape, which may have an impact within the experiments (Rowe 2009). Besides the impact of the process parameters, the impact of the used type of cutting fluid is demonstrated as well. In literature it has already been demonstrated that a high lubrication performance leads to an increase in material displacement induced by ploughing during grinding and thus to an increased grinding power (Winter et al. 2015). As oil-based cutting fluids have better lubricity than emulsions, the power should therefore be higher. This can only be partially confirmed based on the experimental results, as the grinding power is only higher for certain parameter settings when using oil. The experimental results are also not as linear as the theoretical results. This indicates that further extension of the experimental investigations is useful to provide a more precise statement about model quality.

Alternatively to this model, the use of an empirical regression model could lead to less deviation (Winter 2016), but it also means significantly higher effort for the application as regression coefficients need to be determined for each machining system. In contrast, another semi empirical approach based on Kienzle, demonstrated by Meister (2011), is overly simplified and shows greater deviation. For this concept, the presented model (Eq. 4.15–4.19) is selected as it largely meets the requirements and is consistent with turning and milling in terms of required effort.

4.2.2.2 Surface Roughness

In order to characterize and evaluate the quality of the machined workpiece, different parameters can be used. In practice, parameters for the surface roughness are often applied and sufficient in many cases (Klocke 2011). Depending on the different measuring procedures and used filters, various surface parameters are definded in DIN EN ISO 4287 (2010). For this concept, the arithmetic mean roughness value (R_a) is further considered as product-specific indicator, because it has been examined extensively and is often specified for technical designs. This value is defined as the arithmetic mean of the values of the ordinates ($Z(x)$) within a single measuring length (DIN EN ISO 4287 2010):

$$R_a = \frac{1}{L}\int_0^L |Z(x)|dx \tag{4.20}$$

Turning process

Various factors and parameters influence the surface roughness, making it difficult to create a physical model. The model proposed by Ståhl et al. only considers the grooves for the physical calculation of the surface roughness. Within this model, the

R_a-value is theoretically described as a function of the feed rate (f) and the corner radius of the cutting tool (r_ε) as presented by the geometric relationship in Fig. 4.15.

Based on the assumption that the ratio between R_{mean} and R_a is approximately constant at 1.29870 for all feed rates $f < 0.6 \cdot r_\varepsilon$, a model for R_{mean} was first defined and adapted to the geometric correlations to further determine the R_a-value (Ståhl et al. 2011):

$$\begin{aligned} R_{\text{mean}} &= \frac{1}{2r_\varepsilon \cdot \sin|\delta_0|} \int_{\delta_0}^{-\delta_0} r_\varepsilon \cdot (1 - \cos\delta) \cdot r_\varepsilon \cdot \cos\delta d\delta \\ &= r_\varepsilon - \frac{{r_\varepsilon}^2}{f} \arcsin\left(\frac{f}{2r_\varepsilon}\right) - \frac{r_\varepsilon}{2}\sqrt{1 - \frac{f^2}{4{r_\varepsilon}^2}} \end{aligned} \tag{4.21}$$

As the R_a-value describes the deviation of the surface from the centerline along a single measuring length, the integration for Eq. 4.22 was carried out in the direction of the x coordinate and not with respect to the angle δ. Thus, it can be further used to determine the R_a-value (Ståhl et al. 2011):

$$\begin{aligned} R_a &\approx \frac{1}{1.29870} \cdot R_{\text{mean}} \\ &= 0.77 \cdot \left[r_\varepsilon - \frac{{r_\varepsilon}^2}{f} \arcsin\left(\frac{f}{2r_\varepsilon}\right) - \frac{r_\varepsilon}{2}\sqrt{1 - \frac{f^2}{4{r_\varepsilon}^2}} \right] \\ &\approx 0.77 \cdot \left(1 - \frac{\frac{f}{2r_\varepsilon}}{\arcsin\left(\frac{f}{2r_\varepsilon}\right)}\right) \cdot r_\varepsilon \end{aligned} \tag{4.22}$$

This model can be applied given that $f \leq 2\sqrt{a_e(2r_\varepsilon - a_e)} \leq 2r_\varepsilon$ (Ståhl et al. 2011). Following the cutting power models, a correction factor for the cutting fluid (K_{cf}) is also multiplied in the model.

The application of this modeling approach to the case study is demonstrated in Fig. 4.16 for longitudinal turning with a surface measurement inside the workpiece.

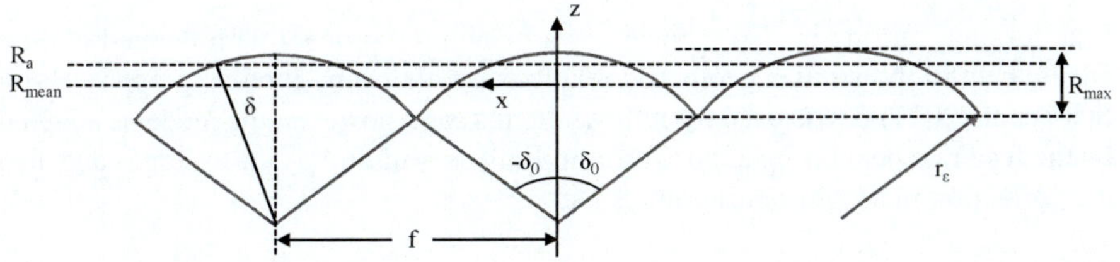

Fig. 4.15 Geometric relationships for the calculation of the surface roughness [according to Ståhl et al. (2011)]

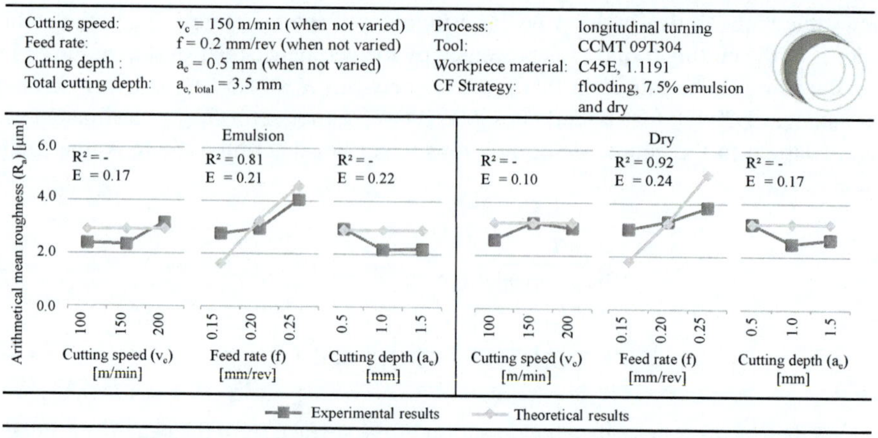

Fig. 4.16 Comparison of modeled and experimental surface roughness of turning processes

In accordance with the model, the experimental results also indicate a linear effect of the feed rate, where an increasing feed rate leads to increasing R_a-values. As the cutting speed as well as the cutting depth are not considered in the model, different parameter settings have no influence on the theoretical results. Regarding the experimental results, variations of these parameters also show effects on the surface roughness, although these are not directly linear. It can therefore be confirmed that especially feed and surface roughness are significantly related. In literature, other studies by e.g. Nalbant et al. (2007) and Sangwan and Sihag (2019) have achieved similar results and revealed that cutting speed and cutting depth are not strongly correlated with surface roughness. Based on the experimental results, the use of cutting fluid demonstrates slightly improved surface quality. In literature, it is assumed that high lubricity prevents materials from sliding directly into one another and thus reduces welding on the surface (Klocke 2011). By comparing the results, a sufficiently low deviation and a good level of accuracy for the feed rate is shown. As cutting speed and cutting depth are not considered in the model, the R^2-values cannot be determined for these parameters (see Eq. 4.1).

Milling process

Like turning, milling is also modeled as a machining process with defined cutting edges. Due to the similar geometric relationship, the same modeling approach by Ståhl et al. (2011) is also used for milling. In this case, however, the model is adapted by the feed rate per cutting edge (f_z), as milling uses multiple cutting edges and also the correction factor for cutting fluids (K_{cf}):

$$R_a = 0.77 \cdot \left[r_\varepsilon - \frac{{r_\varepsilon}^2}{f_z} \arcsin\left(\frac{f_z}{2r_\varepsilon}\right) - \frac{r_\varepsilon}{2}\sqrt{1 - \frac{{f_z}^2}{4{r_\varepsilon}^2}} \right] \cdot K_{cf}$$

$$\approx 0.77 \cdot \left(1 - \frac{\frac{f_z}{2r_\varepsilon}}{\arcsin\left(\frac{f_z}{2r_\varepsilon}\right)}\right) \cdot r_\varepsilon \cdot K_{cf} \tag{4.23}$$

According to the presented procedure, the model is again applied to the case study. In this case, the used milling tool is characterized by a cutting edge width of 0.1 mm at 45°. For simplification, this is used as r_ε-value. The results of the theoretically determined and experimentally measured surface are shown in Fig. 4.17.

Similar to the results for the turning processes, the theoretical surface roughness only indicates an effect of the feed rate. However, this effect could not be verified by the experimental results leading to a larger deviation between the results. The results rather show an increase in surface roughness with increasing cutting depth. Although a similar trend is apparent for the feed rate, this effect would have probably been more significant with greater intervals. Further, the tool shape has been simplified and adapted to fit the model, which could also be a reason for a rather low consistency between experimental and theoretical results.

In literature, further studies focus on modeling the surface roughness, whereby these approaches provide comparable results. The ideal surface roughness for machining processes with defined cutting edges is generally described in literature by the following equation (Boothroyd and Knight 2006):

$$R_a = \frac{0.0321 \cdot f^2}{r_\varepsilon} \tag{4.24}$$

This simplified model is also based on tool shape and feed rate. Further, Wang et al. (2019) developed a model for face milling, which is again based on the geometrical relationship between feed rate and tool shape to determine the R_a-value. This approach uses three different models depending on the ratio of feed rate to cutting

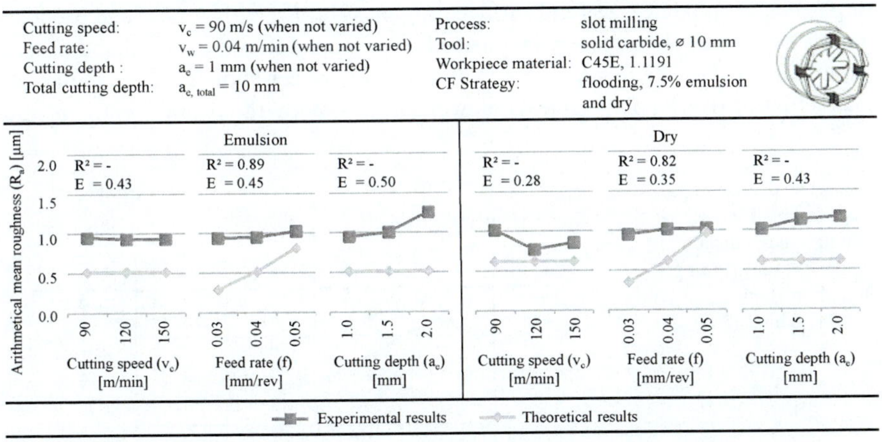

Fig. 4.17 Comparison of modeled and experimental surface roughness of milling processes

edge radius. In comparison, the theoretical R_a-values for the simplified model and the models by Wang et al. are rather similar to the presented results using the approach by Ståhl et al. Wang et al. have also demonstrated that the simplified approach is suitable for low feed rates and provides good accuracy, but should not be used for higher feed rates. As the alternative models do not further improve accuracy and deviation, the model presented in Eq. 4.23 is still selected for the concept. For the outlook, it is recommended to extend the experiments and possibly include further parameters and the workpiece material into the model.

Grinding process

The determination of the surface roughness for grinding differs from the machining processes with defined cutting edges. In this case, multiple undefined cutting edges, more specifically grains, create cutting grooves and piled up material on the groove sides. This process is influenced by the process parameters, but also the characteristics of the grain such as size, shape and material of the grain (Winter 2016). In order to define a simplified model, the piled up material is neglected and only the cutting grooves are considered as presented in Fig. 4.18.

According to Eq. 4.25, the R_a-value can be determined by the areas above and below the center line using the undeformed chip width and thickness:

$$R_a = \frac{4}{b}\int_0^{\frac{b}{4}} \left(\frac{b}{4} - x\right) \cdot \frac{2h_{cu}}{b} dx = 0.25 \cdot h_{cu} \tag{4.25}$$

This proportional relationship between undeformed chip thickness and surface roughness was also revealed by other studies in literature. Khare and Agarwal (2015), however, assumed that that the groove profile resembles an arc of a circle and modeled this by $R_a = 0.471 \cdot h_{cu}$. Hecker and Liang (2003) distinguished between two different types of triangular shaped grooves with a height that is either less or greater than the center line. Overall, Hecker and Liang described the simple relationship between chip thickness and surface roughness by $R_a = 0.37 \cdot h_{cu}$ and proposed the use of empirical correction factors to further adjust it to the considered process requirements. The approach by Su et al. (2019) is rather similar, but considers the interaction of overlapping grooves as well. In this study the overall relationship is

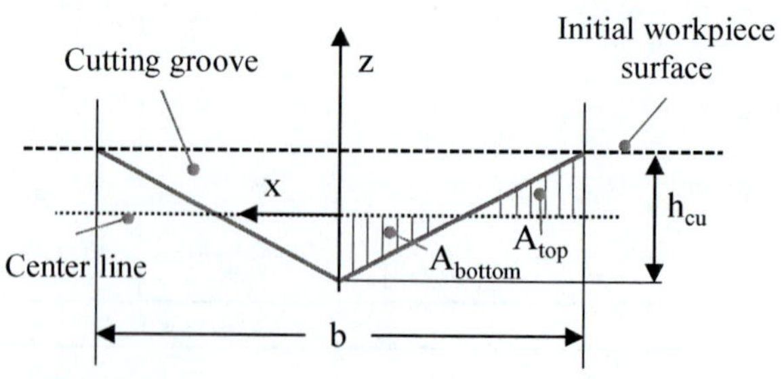

Fig. 4.18 Geometric relationships for the calculation of surface roughness for grinding [based on Hecker and Liang (2003), Winter (2016)]

modeled as $R_a = 0.249 \cdot \sqrt{h_{cu}^2}$. Due to the different constraints, different models are used to determine the thickness in the described approaches.

To remain consistent, Eq. 4.25 is used here for further calculations. The impact of different cutting fluid strategies, however, is not considered within any of the models. Thus, the model used within this work is also extended by an empirical correction factor (K_{cf}) for the impact of the cutting fluid strategy:

$$R_a = 0.25 \cdot h_{cu} \cdot K_{cf} \tag{4.26}$$

Applying this model to the presented case study leads to the following theoretical R_a-values, which are compared to the experimentally measured results in Fig. 4.19.

The results show an improved surface roughness with increasing cutting speed. With increasing workpiece speed and cutting depth, however, the R_a-values increase as well. This theoretical trend is particularly supported by the experimental data for the cutting fluid strategy using oil. It is also conformed by fairly high R^2-values. Overall, the use of a water-based emulsion leads to a higher surface roughness compared to the use of oil. This relationship between lubrication performance and surface roughness was also demonstrated in literature. It is assumed that a low lubricity causes breaking and splintering of the grains, resulting in deeper grooves and thus worse surface roughnesses (Winter et al. 2015).

Although a simplified model based on a proportional correlation between surface roughness and chip thickness is used here, the results are sufficiently accurate and meet the requirements of the concept.

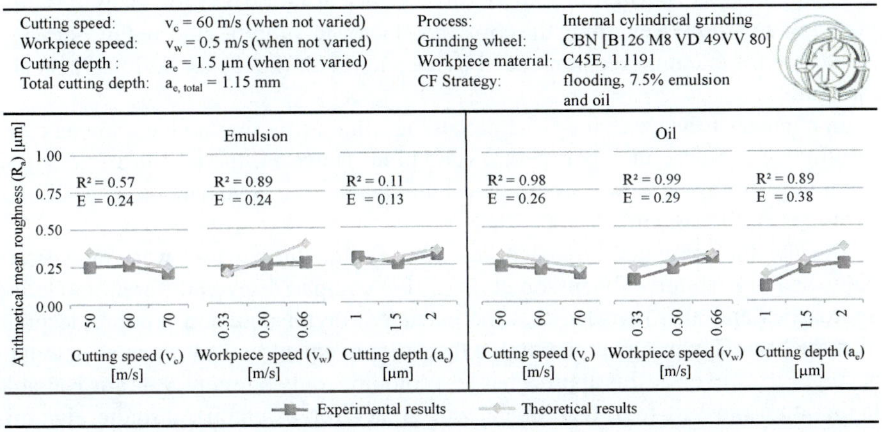

Fig. 4.19 Comparison of modeled and experimental surface roughness of grinding processes

4.2.2.3 Tool Wear

According to the concept, the wear of the used tool is analyzed. Based on this wear, the tool life is determined and thus the number of required tools or inserts can be taken into account for the process design. In case of grinding, the wear also indicates the need for a next dressing cycle.

Turning process

The first wear models were based on the research by Taylor from the early 1900s. Within this approach, an exponential relationship between cutting speed and tool life was determined by experimental investigations, which is referred to as the Taylor equation. As this model focuses only on the impact of cutting speed on tool life, various investigations were performed by e.g. Kronenberg (1969) to improve the accuracy of the model. Thus, the influence of feed rate as well as cutting depth are also considered within the extended Taylor equation. This model also requires additional constants. However, the influence of different cutting fluid strategies are not taken into account by these models. The additional correction factor (K_{cf}) is therefore added in line with the previous models. The tool life (t_l) is determined as follows:

$$t_l = \mathrm{C} \cdot v_c^k \cdot f^{k_f} \cdot a_e^{k_a} \cdot K_{cf} \tag{4.27}$$

The constants C, k, k_f and k_a differ depending on the used tool and material. Similar to the cutting force model, values for the constants are available in the literature such as by Dietrich (2016) and Degner et al. (2015) for many different materials and tools. Regarding the correction factor K_{cf} for the cutting fluid strategy, values are for example recommended by Degner et al. (2015) for various tools and materials.

In context of the case study, the processing time using the same parameters was insufficient to make an experimental validation. However, the tool life calculation according to Taylor is widely used in literature by e.g. (Madanchi et al. 2019b; Schlosser 2013; Songmene et al. 2018).

Besides the presented approach based on Taylor, further tool wear models are published in literature. Johansson et al. (2017) compared seven different modeling approaches and also revealed that the extended Taylor equation is more accurate than the basic Taylor equation but that the Colding equation leads to an even higher accuracy. However, as this approach is less common, only few constants are available in literature and have to be determined with great effort by empirical studies. For this reason, the extended Taylor equation is considered in the context of this concept.

Milling process

Regarding machining processes with geometrically defined cutting edges, the previous results already demonstrated a certain transferability of the theoretical models from turning to milling. Consequently, the extended Taylor equation is also used to model the tool wear for milling processes. In the case of milling, however, the number of used cutting edges has to be modeled as well (Degner et al. 2015):

$$t_l = C \cdot v_c^k \cdot f^{k_f} \cdot a_e^{k_a} \cdot \left(\frac{z}{D}\right)^{k_z} \cdot \left(\frac{a_p}{D}\right)^{k_{ae}} \cdot K_{cf} \tag{4.28}$$

This model considers the impact of multiple cutting edges (z_c) as well as the cutting width (a_p) and contains the additional constants (k_z) and (k_{ae}). Similar to turning, experimental validation could also not be performed during the case study.

Grinding process

In view of the structure of a grinding wheel and its varying wear mechanisms, a different modeling approach has to be applied for grinding compared to turning and milling. In this case, grains splitter and break out of the bond (Marinescu et al. 2007). A model for the analytical determination of the grinding wheel wear volume (V_{gt}) was derived by Winter (2016) based on Archard (1961):

$$V_{gt} = K_{c,g} \cdot F_n \cdot l_c \tag{4.29}$$

It considers the cutting length (l_c), which is determined by the geometrical contact length (l_g), cutting speed (v_c), diameter of the grinding wheel (d_g) and processing time (t_c):

$$l_c = l_g \cdot \frac{v_c}{\pi \cdot d_g} \cdot t_c \tag{4.30}$$

The abrasive wear coefficient of the grain ($K_{c,g}$) is calculated according to Torrance and Badger (2000):

$$K_{c,g} = 0,003 \cdot \frac{27 \cdot (1,185 \cdot \tan\alpha)^3}{f_{cf} \cdot H_{s,g} \cdot \sqrt{l}} \cdot \sqrt{\frac{H_{b,g}}{H_{s,g}}} \tag{4.31}$$

It includes grain surface hardness ($H_{s,g}$), bulk material hardness ($H_{b,g}$) as well as the dimensionless distance between adjacent tracks (l). The interfacial friction coefficient (f_{cf}) between the cutting fluid and the grinding wheel depends on the type of cutting fluid. Further, the attack angle (α) is considered, which is calculated by the following equation using the grain tip radius (r_g) and chip thickness:

$$\alpha = \arccos\left(\frac{r_g - h_{cu}}{r_g}\right) \tag{4.32}$$

Further, the normal force is required to determine the wear volume, which can be calculated as follows:

$$F_n = \frac{P_c}{\mu \cdot v_c} \tag{4.33}$$

According to Xie and Williams (1996), the required friction coefficient (μ) is formulated as:

$$\mu = \left(\frac{2}{\pi}\right)^{\frac{1}{2}} \cdot \tan\alpha \cdot \left(1 + f_{cf} \cdot \left(1 + \frac{\pi}{4 \cdot \tan^2\alpha}\right)^{\frac{1}{2}}\right) \tag{4.34}$$

This model considers the impact of the used cutting fluid through the friction coefficient. It is generally assumed that the higher lubricating effect of the oil results in less wear, as the grains rather blunt and do not splinter or break (Winter et al. 2015). Similar to turning and milling, the experimental validation is not performed.

4.2.3 Peripheral Subsystem Model

According to the concept, it is further required to model the different peripheral subsystems. While the machining process models are usually dependent on the type of process e.g. turning, milling or grinding, the models of the peripheral subsystem are generally valid and independent from the type of machining process.

4.2.3.1 Cutting Fluid System

As presented in Sect. 2.3.2, the cutting fluid system consists of different subsystems and components that are modeled separately. Their relevance and influence on energy and resource efficiency was demonstrated in that section as well. Following this, theoretical models are combined to an integrated cutting fluid system containing these components as demonstrated in Fig. 4.20.

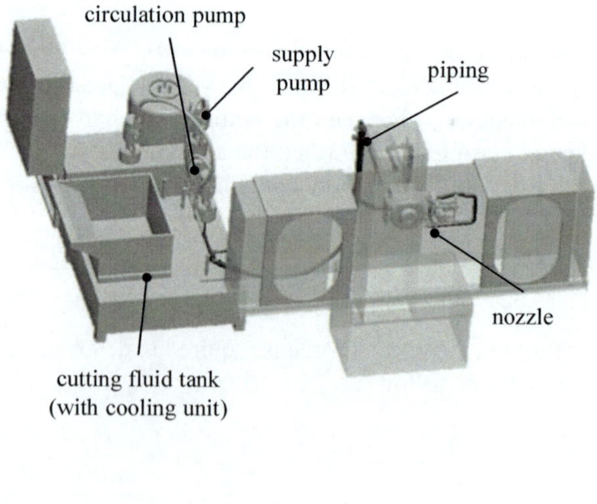

Fig. 4.20 Components and subsystems of the cutting fluid system

Piping

The piping causes pressure losses due to friction on the inner walls of the pipes and due to local losses in valves or fittings. Based on general fluid mechanical relationships, these pressure losses can be calculated (Sigloch 2008). Thus, the pressure loss due to pipe friction (Δp_f) is determined as follows:

$$\Delta p_f = \lambda \cdot \frac{l_p}{d_i} \cdot \frac{\rho}{2} \cdot v^2 \tag{4.35}$$

In this case, the pressure loss depends on the length (l_p) and inner diameter (d_i) of the pipe as well as flow velocity (v) and friction factor (λ). The determination of the friction factor, however, depends on the Reynolds number (Re):

$$Re = \frac{v \cdot d_i}{\nu_{cf}} \tag{4.36}$$

In order to calculate the Reynolds number, the kinematic viscosity of the cutting fluid (ν_{cf}) is also critical. It describes whether the flow is laminar or turbulent and thus determines the calculation of the friction factor. In the case of laminar flow with $Re < 2320$, Eq. 4.37 is applied:

$$\lambda = \frac{64}{Re} \tag{4.37}$$

However, in case of a turbulent flow with Re > 2320, the calculation also depends on the relative wall roughness (k_s). With increasing wall roughness the equations from (4.38) to (4.40) should therefore be applied.

Constraint	Equation	
$\frac{k_s}{d_i} \cdot Re^{0,875} < 25$	$\lambda = \frac{0,316}{Re^{0,25}}$	(4.38)
$25 \leq \frac{k_s}{d_i} \cdot Re^{0,875} < 350$	$\frac{1}{\sqrt{\lambda}} = -2 \cdot lg\left(\frac{2,51}{Re \cdot \sqrt{\lambda}} + 0,27\frac{k_s}{d_i}\right)$	(4.39)
$\frac{k_s}{d_i} \cdot Re^{0,875} \geq 350$	$\lambda = \frac{1}{\left(2 \cdot lg\frac{d_i}{k_s} + 1,14\right)^2}$	(4.40)

For planning a new machining system, the design and selection of technical components is still more flexible. This provides the possibility to determine the inside diameter of the pipe on the basis of supplied cutting fluid volume flow ($\dot{V}_{cf}$) and an economic flow velocity (v_{eco}) as presented in Eq. (4.41). Possible values for the flow velocity are recommended, for example, by Triesch (2010) for the use of oil and emulsions (see Sect. 2.3.2.1).

$$d_i = \sqrt{\frac{\dot{V}_{cf} \cdot 4 \cdot 10^3}{v_{eco} \cdot \pi \cdot 60}} \tag{4.41}$$

Additionally, installations such as valves and pipe transitions in the piping system must also be defined and examined as they lead to local pressure losses (Δp_i):

$$\Delta p_i = \varsigma \cdot \frac{\rho_{cf}}{2} \cdot v^2 \tag{4.42}$$

Besides the cutting fluid density (ρ_{cf}) and flow velocity (v), it particularly depends on the loss coefficients (ς), which are assigned to a variety of different pipe elements and installations.

From these two possibilities of pressure losses it is then necessary to determine the total pressure head loss (H_l), which has to be compensated by an increased pump power:

$$H_l = \frac{p_{cf,l}}{\rho_{cf} \cdot g}$$

$$\sum H_l = \frac{\sum_{e=1}^{E} \Delta p_{f,e} + \sum_{n=1}^{N} \Delta p_{i,n}}{\rho_{cf} \cdot g} \tag{4.43}$$

For this purpose, the pressure loss for each pipe element (e) with a specific friction and for each pipe installation or transition (n) has to be summed up.

Nozzle

At the end of the piping system, the cutting fluid is released through a nozzle. The resulting cutting fluid volume flow ($\dot{V}_{cf}$) is especially relevant and varies depending on the used machining process. For grinding, Astakhov and Godlevskiy (2012) recommend at least a volume flow of 0.8 l/min per mm grinding wheel width, whereas Marinescu et al. (2007) recommend 5.7–7.5 l/min per kW/h. Regarding the machining processes with geometrically defined cutting edges, a cutting fluid volume flow of 8–20 l/min is recommended for turning, a value of 15–20 l/min for milling and 8–15 l/min for drilling (Tornau et al. 1999). However, Denkena et al. (2014) demonstrated within a specific case study that the optimal cutting fluid volume flow for these machining processes can also be less than 6 l/min. In addition to these recommendations, Metzger (1986) proposed an approach to model the minimal required cutting fluid volume flow depending on the grinding power, specific heat capacity ($c_{p,cf}$) and density of the fluid (ρ_{cf}), tolerable temperature rise ($\Delta T_{cf,to}$) and supply efficiency (η_{CF}). This approach is also used by Winter (2016):

$$\dot{V}_{cf} = \frac{P_c}{c_{p,cf} \cdot \rho_{cf} \cdot \Delta T_{cf,to} \cdot \eta_{CF}} \tag{4.44}$$

This model is based on the approach to dissipate the heat generated, while taking into account the fluid properties. As the focus here is on the general cooling function of cutting fluids, the approach can also be transferred from grinding to turning and milling.

In order to determine the resulting volume flow at the nozzle outlet, the continuity equation is applied, which describes the relationship between volume flow ($\dot{V}_{cf}$), flow velocity (v) and outlet area (A_n) (Sangermann 2013):

$$\dot{V}_{cf} = v \cdot A_n = \text{constant} \tag{4.45}$$

In case of grinding, the flow velocity at the nozzle outlet is recommended to be equal or slightly less than the grinding wheel speed (Morgan et al. 2008). However, for other machining processes, the resulting jet pressure is more relevant and is determined by the jet force (F_j) and area (A_j):

$$p_j = \frac{F_j}{A_j} \tag{4.46}$$

Assuming that the jet area is equal to the nozzle outlet area and thus the expansion can be neglected, the volume flow is calculated as follows (Sangermann 2013):

$$\dot{V}_{cf} = \sqrt{\frac{p_j}{\rho_{cf}}} \cdot A_{j/n} \tag{4.47}$$

At the nozzle outlet, there can be flow losses and thus a narrowing of the jet, resulting in a lower volume flow. The ratio between actual and calculated volume flow is described by the flow coefficient. For well-rounded nozzles, this coefficient is assumed to be between 0.97 and 0.99 (Brücher 1996; Sigloch 2008), and will further be neglected under this assumption.

Pump

With regard to cutting fluid systems, pumps are particularly critical. They are used for the supply of the cutting fluid to the machine tool, but also to circulate it within the filtration system or to return the fluid from machine tool to the cutting fluid system by pump back stations.

Based on general fluid mechanical relationships, the mechanical pump power ($P_{P,m}$) can be determined as follows:

$$P_{P,m} = \frac{\dot{V}_{cf,P} \cdot H_P \cdot \rho_{cf} \cdot g}{\eta_{hyd}} \tag{4.48}$$

It therefore depends on the cutting fluid volume flow at the pump ($\dot{V}_{cf,P}$), the pump head (H_P), cutting fluid density (ρ_{cf}), gravitational constant (g) and hydraulic efficiency (η_{hyd}). The possible pump head is described by the pump performance curve and is thus decisive for the pump selection. By using a VFD to control the supply of centrifugal or screw pumps, the operating point can be set directly. In this case, the pump head (H_P) equals the system head (H_{sys}), which is determined by using Bernoulli's equation and Eq. 4.43:

$$H_{sys} = (h_2 - h_1) + \frac{p_2 - p_1}{\rho \cdot g} + \frac{v_2^2 - v_1^2}{2 \cdot g} + \sum H_l \quad (4.49)$$

On the assumption of an open tank, the pressure head difference $(p_2 - p_1)$ is not taken into account (Kuhrke 2011). Further, the initial flow velocity v_1 starting from the tank is 0 and can also be neglected. By using a throttle instead of a VFD for the centrifugal pump, the required pump head has to be determined from the pump performance curve.

The hydraulic efficiency is usually provided by the pump manufacturer for a fluid with the kinematic viscosity of $\nu_A = 1 \cdot 10^{-6}$ m²/s. In case of centrifugal pumps, the hydraulic efficiency highly depends on the viscosity and has to be adapted to the specific cutting fluid viscosity by the following equation (Pfleiderer 1961):

$$\eta_B = 1 - (1 - \eta_A) \cdot \left(\frac{\nu_B}{\nu_A}\right)^{0,1} \quad (4.50)$$

Here, the viscosity (ν_A) equals to $1 \cdot 10^{-6}$ m²/s and the viscosity (ν_B) represents the cutting fluid viscosity (ν_{cf}).

To determine the electrical power of the pump motor $(\mathrm{P}_{P,el})$, Eq. 4.48 is extended by the motor efficiency (η_{mot}):

$$\mathrm{P}_{P,el} = \frac{\dot{\mathrm{V}}_{cf,P} \cdot H_P \cdot \rho_{cf} \cdot g}{\eta_{hyd} \cdot \eta_{mot}} \quad (4.51)$$

The efficiency can be calculated using the approach defined by DIN EN 60034-30-1 (2014) and depends on the rated motor power as well as the energy efficiency classes (IE1–IE4) (see Sect. 2.3.2.1). To transfer it to this approach, however, the electrical power of the pump motor is used and also applied at partial load.

$$\eta_{mot} = A \cdot \left[\log_{10}\frac{\mathrm{P}_{P,m}}{1\mathrm{kW}}\right]^3 + B \cdot \left[\log_{10}\frac{\mathrm{P}_{P,m}}{1\mathrm{kW}}\right]^2 + C \cdot \left[\log_{10}\frac{\mathrm{P}_{P,m}}{1\mathrm{kW}}\right] + D \quad (4.52)$$

The coefficients (A, B, C, D) for each efficiency class can be taken from the standard and are valid for a power range starting from 0.12 kW (DIN EN 60034-30-1 2014). Further, when using a VFD, its efficiency must also be taken into account. It is assumed that the rated VFD power equals the maximal electrical pump motor power, which is at the operation point with the highest volume flow for centrifugal pumps and with the highest pump head for screw pumps. The values for the VFD efficiency are taken from the standard, which describe different efficiency curves (DIN CLC/TS 60034–31 2011). The values between the curves are linearly interpolated. Therefore, when using an inverter, the electrical power of the pump motor $(\mathrm{P}_{P,el})$ is calculated according to the following equation:

$$P_{P,el} = \frac{\dot{V}_{cf,P} \cdot H_P \cdot \rho_{cf} \cdot g}{\eta_{hyd} \cdot \eta_{mot} \cdot \eta_{VFD}} \tag{4.53}$$

Cooling

In order to maintain the cutting fluid quality, it is required to keep its temperature plus or minus two to three degrees of the ambient temperature (Byers 2006). However, as described in Sect. 2.3.2.3, there are different heat inputs into the cutting fluid. This is the heat input due to the pumps ($\dot{Q}_{cf,p}$), which involves all filtration and circulation pumps as well as supply pumps of the system (m). Further, the heat input converted from the mechanical cutting energy from the cutting process ($\dot{Q}_{cf,c}$) has to be taken into account as well. For a centralized system it is important to consider the machining process of each machine tool (mt) within the process chain. However, heat can also be also released into the environment through pipes and tanks. The heat losses due to radiation ($\dot{Q}_{cf,r}$) and convection ($\dot{Q}_{cf,con}$) are summed up for each component (c). This results in an approximate heat flow that has to be removed by the cooling unit ($\dot{Q}_{cf,cu}$) and is determined according to the following equation:

$$\dot{Q}_{cf,cu} = \sum_{m=1}^{M} \dot{Q}_{cf,p,m} + \sum_{mt=1}^{MT} \dot{Q}_{cf,c,mt} - \sum_{c=1}^{C} \dot{Q}_{cf,r,c} - \sum_{c=1}^{C} \dot{Q}_{cf,con,c} \tag{4.54}$$

To determine the heat input by the pumps, there is a direct relation to its power consumption. It can be assumed that the entire power loss is converted into heat and therefore a calculation can be made on the efficiency by using Eq. 4.50 and Eq. 4.52 (Züst 2017):

$$\dot{Q}_{cf,p} = P_{P,el} \cdot (1 - \eta_{mot}) \cdot (1 - \eta_{hyd}) \tag{4.55}$$

As a more simplified approach, Triebs (2015) assumed that 40% of the electrical power consumption of a pump is liberated as heat into the cutting fluid based on pump producers and experiments, while a case study by Herrmann et al. (2009) demonstrated a value of about 24%. Similarly, it is assumed that 66–87% of the power consumption resulting from the machining process is also liberated as heat into the cutting fluid (Knobloch 1979). For simplification, an average value of 76.5% is taken into account for this concept:

$$\dot{Q}_{cf,c} = P_c \cdot 0.765 \tag{4.56}$$

The heat losses due to heat radiation and convection are determined based on physical relations (VDI 2055 2019). Heat radiation is an electromagnetic radiation, which is emitted and absorbed by solids, liquids or gases. The heat flow depends on the absolute temperatures of both components ($T_{1,2}$), the surface area (A_i) and the radiation coefficient ($C_{r,1,2}$):

$$\dot{Q}_{cf,r} = C_{r,1,2} \cdot \left(T_1^4 - T_2^4\right) \cdot A_i \tag{4.57}$$

For pipes or tanks that are indoors, it is possible to simplify the radiation coefficient by using the Stefan-Boltzmann-constant (σ) and the emission level of the surface (ε_1):

$$C_{r,1,2} = \varepsilon_1 \cdot \sigma \tag{4.58}$$

The heat loss due to heat convection is described as a transport mechanism in liquids and gases, in which a heated surface transfers heat to a medium flowing around it. For the cutting fluid system, this medium is the ambient air in the factory building and it is a free convection without external influences such as ventilators. This heat flow is determined based on the temperature difference between cutting fluid system and ambient air ($\Delta T_{1,2}$), surface area (A_i) and the convective heat transfer coefficients (α_c):

$$\dot{Q}_{cf,con} = \alpha_c \cdot \Delta T_{1,2} \cdot A_i \tag{4.59}$$

The convective heat transfer coefficients can further be formulated as a function of the dimensionless Nusselt number (Nu), flow length (l_f) and heat conductivity of air (λ_{air}):

$$\alpha_c = \frac{\lambda_{air} \cdot Nu}{l_f} \tag{4.60}$$

The Nusselt number further has to be determined by the Grashof number (Gr) and Reynolds number (Re):

$$Gr = \frac{g \cdot {l_f}^3 \cdot \left|\Delta T_{1,2}\right|}{{\upsilon_{air}}^2 \cdot T_{air}} \tag{4.61}$$

$$Re = \frac{v \cdot l_f}{\upsilon_{air}} \tag{4.62}$$

Values for the heat conductivity and kinematic viscosity of the air (υ_{air}) are temperature dependent constants and can be taken from literature. The calculation of the Nusselt number further depends on the type and orientation of the component and is formulated as follows for a vertical pipe:

$$Nu = (0.752 + 0.303 \cdot Gr^{\frac{1}{6}})^2 \tag{4.63}$$

The actual temperature increase of the cutting fluid, however, is time dependent and further considers the cutting fluid density (ρ_{cf}), volume flow ($\dot{V}_{cf}$) and specific heat capacity of the cutting fluid ($c_{p,cf}$):

$$\Delta T_{cf(t)} = \frac{\dot{Q}_{cf,c(t)}}{\dot{V}_{cf}\cdot\rho_{cf}\cdot c_{p,cf}} \tag{4.64}$$

Following the determination of the heat flow it is required to derive the required cooling power from it. It is further based on the energy efficiency ratio (EER), which describes the ratio of cooling power to the consumed electrical power of a cooling unit (P_{cu}) (DIN EN 14511-1 2019):

$$P_{cu} = \frac{\dot{Q}_{cf,cu}}{EER} \tag{4.65}$$

Similar to the machining process models, the presented theoretical models regarding the cutting fluid system are also applied and verified experimentally. In this case, the supply system of a grinding machine is considered as presented in Fig. 4.21. It consists of a filtration unit, pumps, piping system with installations such as valves and sensors and a nozzle.

The results focus on the electrical power of the supply pump (5) including pressure loss, the volume flow through the nozzle (9) and the cutting fluid temperature (3). The

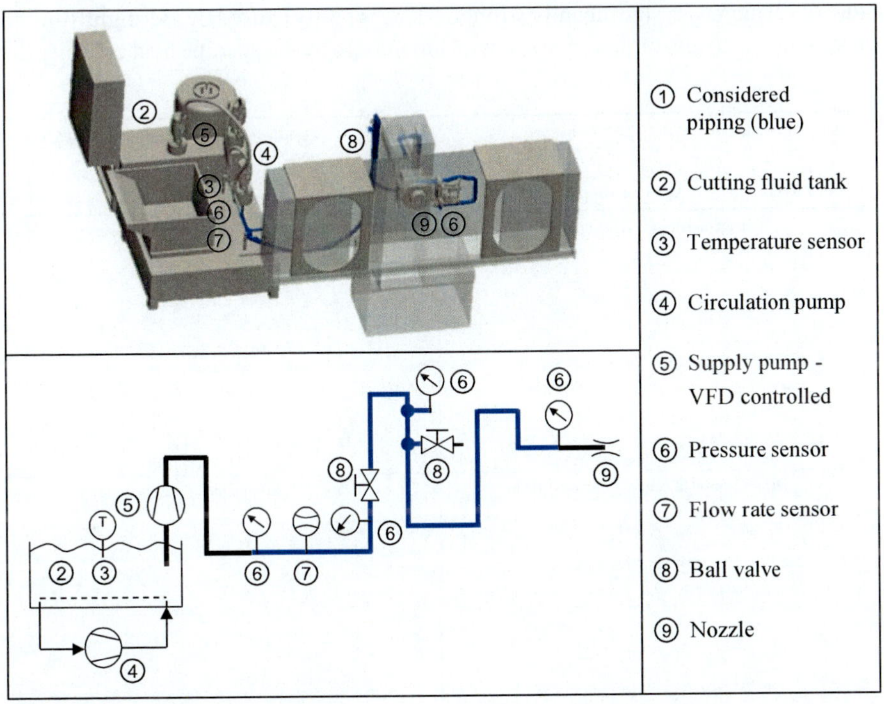

Fig. 4.21 Cutting fluid supply system[1]

[1] The piping system does not show accurate dimensioning.

experimentally observed and theoretically determined results are shown in Fig. 4.22. For the use of oil, experimental data are only presented with regard to temperature development, the other experimental data refer to emulsion.

In order to determine the theoretical electrical pump power, Eqs. 4.48–4.53 are used. This takes into account the pressure loss through the piping system, which is between 0.5 to 2 bar depending on the volume flow rate. As the pumps are controlled by a VFD, it can be assumed here that the pump head (H_P) is equivalent to the system head (H_{sys}). The results show a similar course with slightly higher theoretical results for the use of oil than emulsion. This difference is caused by the difference in fluid densities. Comparing the theoretical and experimental results for emulsion, a high level of consistency with small deviations can be observed.

Further, the resulting flow rate at the nozzle outlet is investigated in relation to different pressure levels. The theoretical results show a similar increase as the experimental results and thus a high degree of consistency. The offset between the used cutting fluid strategies is again due to the different fluid densities, as it also included in the models for theoretical determination.

The development of the cutting fluid temperature is presented in the lower diagrams of the figure. The left diagram shows the experimentally measured temperature curve for using oil compared to emulsion. It is clearly illustrated that the temperature of oil increases significantly stronger. This is partly caused by the slightly higher heat input due to e.g. higher pump power, but mainly by the specific heat capacity that

Pump pressure:	p_P = 1.47 – 8.46 bar	Pump:	Centrifugal pump; Brinkmann TC250/520
Volume flow rate:	V_{cf} = 14.4 – 35.6 l/min	CF strategy:	flooding, 7.5% emulsion and oil
Pipe diameter:	d_i = 25 – 30 mm	Kin. viscosity:	υ_{cf} = 0.87 mm²/s (emulsion); 12 mm²/s (oil)
Nozzle outlet:	A_n = 16 mm²	Density:	ρ_{cf} = 1.01 g/cm³ (emulsion); 0.89 g/cm³ (oil)

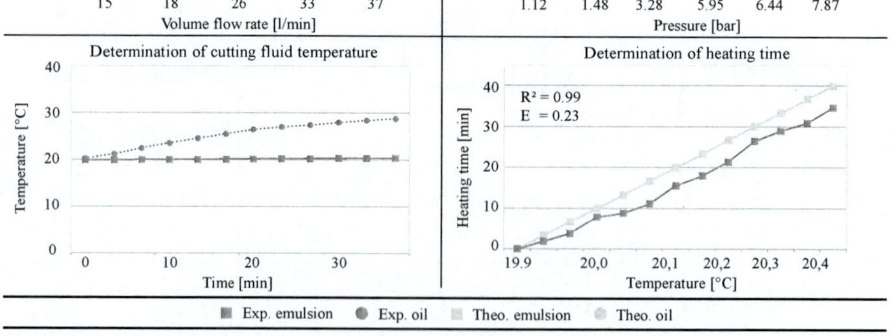

Fig. 4.22 Comparison of modeled and experimental cutting fluid supply system

differs for each fluid. Using Eq. 4.64 the heating time is theoretically determined to increase the cutting fluid temperature analogous to the experiment. The results for the use of emulsion are demonstrated in the right diagram and show a high consistency with a sufficiently low relative error between the results.

4.2.3.2 Exhaust Air System

The exhaust air system is required to remove the cutting fluid emissions from the machine tool in order to ensure occupational health and safety as described in Sect. 2.3.3. Usually, the exhaust air system is operated at a fixed state and removes a constant air flow, which is independent of the supplied cutting fluid quantity (Winter 2016). Regarding this subsystem, the modeling of the electric fan power is particularly relevant and can be determined as follows (DIN EN 16798–3 2017):

$$P_{ea} = \frac{\Delta p_V \cdot \dot{V}_{ea}}{1000 \cdot \eta_{ea}} \tag{4.66}$$

The fan power (P_{ea}) thus depends on the air flow ($\dot{V}_{ea}$), the total pressure increase (Δp_V) as well as the efficiency (η_{ea}). The required air flow highly depends on the size of the workstation volume (V_{ws}), the used cutting fluid strategy, the type of machining process as well as the intensity of machining and quality of extracted air ducts (VDI 3802-2 2012). Based on the VDI guidelines a simplified model was derived, which assumes a medium intensity and quality:

$$\dot{V}_{ea} = a_{ea} + b_{ea} \cdot V_{ws} \tag{4.67}$$

The coefficients (a_{ea}, b_{ea}) are presented in the appendix for different cutting fluid strategies and machining processes. This approach to determine the electrical power was already used by Madanchi et al. (2019b) and compared to experimental results with a high level of agreement.

Besides the power consumption, the exhaust air system is further relevant regarding the drag-out of cutting fluid and thus the consumption of resources. As discussed in Sect. 2.3.3.1, the generation of cutting fluid emissions highly depends on the selected process parameters such as cutting speed and volume flow. In the context of the research study by Madanchi et al. (2015a), the drag-out is modeled for a grinding process using this empirical nonlinear regression model:

$$\dot{m}_{cf,ea,g} = K_{cf,ea} + C_{ea,g} \cdot \dot{V}_{cf}^{\alpha_1} \cdot v_c^{\alpha_2} \cdot \dot{V}_{ea}^{\alpha_3} \cdot \upsilon_{cf}^{\alpha_4} \cdot \sigma_{cf}^{\alpha_5} \tag{4.68}$$

It contains the cutting fluid properties surface tension (σ_{cf}) and viscosity (υ_{cf}) as well as the process parameters cutting speed (v_c) and flow rate of the cutting fluid ($\dot{V}_{cf}$) and air ($\dot{V}_{ae}$). Following this modeling approach, the drag-out of cutting fluid emissions is also investigated for turning processes. However, the measuring procedure to experimentally determine the drag-out differs for both processes, as an airborne particle counter is used for turning (see Sect. 4.2.1.2) and a filter paper based procedure for grinding (see Madanchi et al. (2015a)). The absolute values can therefore not be compared directly. As the diameter of the rotating workpiece has also an impact, the rotational speed (n_c) is considered for turning instead of the cutting speed. Further, it is revealed that the process parameters feed rate (f) and cutting depth (a_e) are also significant, whereas the surface tension is not.

$$\dot{m}_{cf,ea,t} = K_{cf,ea} + \dot{V}_{cf}^{\beta_1} \cdot n_c^{\beta_2} \cdot \dot{V}_{ea}^{\beta_3} \cdot f^4 \cdot a_e^{\beta_5} \tag{4.69}$$

The derived regression coefficients for different cutting fluid strategies and the machining processes are presented in the appendix.

The results for the theoretically and experimentally determined drag-out is shown for grinding in Fig. 4.23. It is striking that the drag-out for the used oil is significantly higher compared to the use of an emulsion. Regarding the influencing parameters, the cutting speed demonstrates the highest impact on the generation of emissions and thus the drag-out. This is in line with the results from literature. While the exhaust air flow rate has nearly no influence when using emulsion, it shows a very clear effect for using oil. The increasing cutting fluid volume flow shows the same increasing impact on the drag-out for both fluids.

The results of turning, however, differ slightly from this (see Fig. 4.24). In case of turning, the flow rate of the exhaust air, rotational speed and feed rate demonstrate the most significant impact. The increasing volume flow of the cutting fluid and cutting depth only lead to a slightly increasing trend on drag-out and thus have less impact. Based on Eq. 4.67 and the VDI guidelines, the exhaust air system should be operated at a flow rate of around 13.8 m^3/min for using oil and 11.2 m^3/min for using emulsion. This flow rate does not have a direct impact on the generation of cutting fluid emissions, but on the drag-out. Larger particles, which would otherwise have

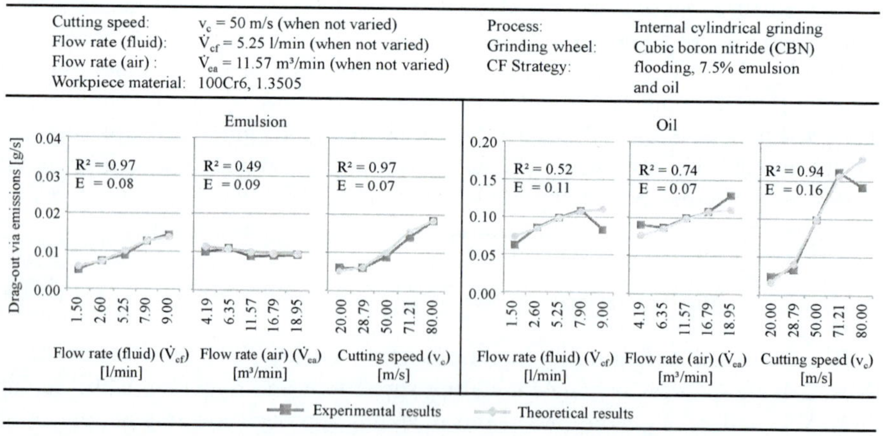

Fig. 4.23 Comparison of modeled and experimental drag-out via emissions - grinding processes

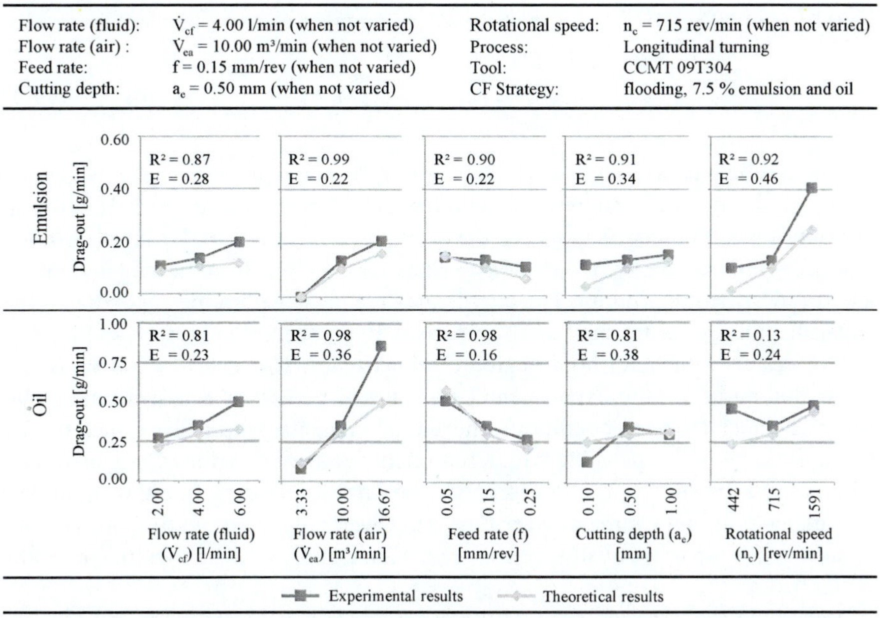

Fig. 4.24 Comparison of modeled and experimental drag-out via emissions - turning processes

settled again, could be unnecessarily removed and dragged out, if the volume flow is oversized. Similar to the results in grinding, the use of oil shows a significantly higher formation of emissions and also drag-out compared to the use of emulsion.

Overall, the results demonstrate high R^2-values and sufficiently low E-values for both processes, which indicates a high agreement between the theoretical and experimental results. However, due to the use of empirical models the determined regression coefficients cannot be fully transferred to other processes and machining systems. In order to avoid additional effort to re-determine the coefficients, future research should focus on the use of a physical model.

4.2.3.3 Workpiece and Scrap Treatment System

Additional to the drag-out via the exhaust air system, cutting fluid drag-out can also result from wetting the workpiece and scrap material. According to Winter (2016), this accounts for the largest share. The wetting highly depends on the shape and size of the workpiece surface and the viscosity of the fluid. Low viscosity cutting fluids tend to drip more easily from the workpiece, while higher viscosity fluids adhere to the workpiece surface (Dettmer 2006). In order to determine the amount of cutting fluid adhering to materials, an empirical approach is used here as well, because a physical and generally valid model could not be identified for it. The empirical regression model proposed by Petuelli (2002) determines the amount of adhered cutting fluid to workpieces and scrap ($\dot{m}_{cf,ws}$) over time (t_{ws}) and was also used by Winter (2016):

$$\dot{m}_{cf,ws} = a_{ws} + \frac{b_{ws}}{t_{ws}} + c_{ws} \cdot \sqrt{t_{ws}}\ for\ 3\,\mathrm{s} < t_{ws} < 3600\,\mathrm{s} \tag{4.70}$$

The coefficients (a_{ws}, b_{ws}, c_{ws}) were derived from simple investigations in which workpieces and chips were wetted and the weight difference was measured over the dripping time. The coefficients for the workpiece from Fig. 4.4 were determined after different machining processes as well as for the chips from turning and milling and the swarf from grinding. The coefficients are presented in the appendix and the experimental results are shown in comparison to the theoretical results in Fig. 4.25.

The results indicate a high degree of consistency. However, as the model is a regression based on the experimental data, it can be assumed that it will be less accurate under changed conditions and the coefficients need to be re-determined. For this model this is possible with a reasonable amount of effort. Generally, it can be observed that within the first seconds the largest amount of cutting fluid drips off and the drag-out decreases over time. However, after a certain time the drag-out tends towards a constant value, so the regression is only valid up to a time of 3600 s (Petuelli 2002).

Regarding the wetting of the workpiece, the use of oil after the milling process leads to the highest drag-out. This might be due to the significantly higher viscosity

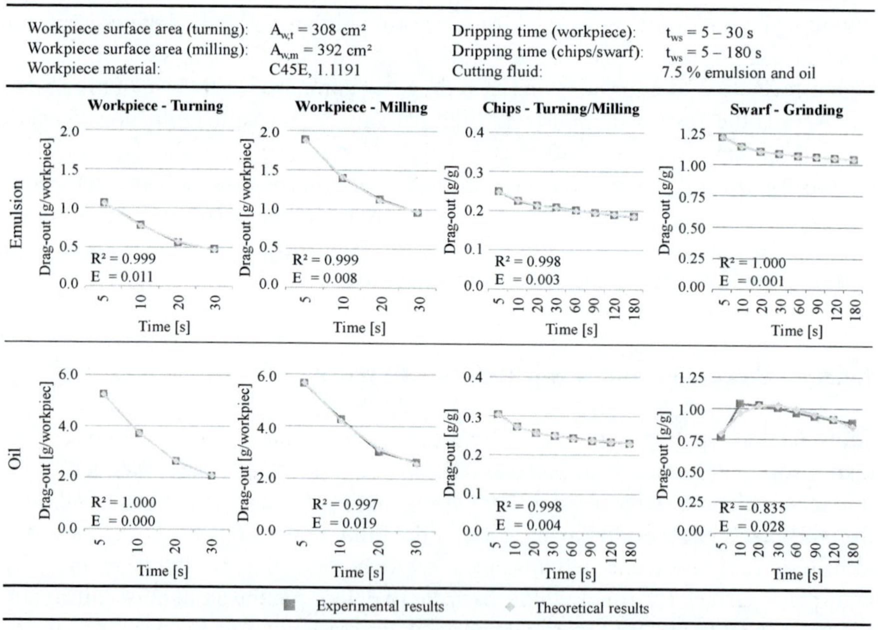

Fig. 4.25 Comparison of modeled and experimental drag-out via wetting of workpieces, chips and swarf

of the oil compared to the emulsion. As a result, the oil has a lower tendency to flow and adheres more to the surface. Further, the surface area and scooping elements increased after milling, which led to a higher drag-out. As the geometry basically remained the same after grinding, the workpiece was not examined separately. With regard to the chips, the oil again demonstrated a higher drag-out, but this was not the case for the swarf. The higher wetting of the emulsion in this case might be due to the used filtration material, which was also wetted by the emulsion and prevented it from passing through. A higher drag-out for the emulsion regarding the filtration material was also observed by Winter (2016). Overall, the drag-out via the swarf is significantly higher compared to the chips and also a little higher compared to other studies presented in Sect. 2.3.4.2.

In general, it is required to determine the cutting fluid drag-out for a specific number of workpieces (n_{ws}) with a defined surface area (A_{ws}):

$$m_{cf,w} = \dot{m}_{cf,w} \cdot \frac{A_{ws} \cdot n_{ws}}{\rho_{cf}} \tag{4.71}$$

For the drag-out via scrap material, the volume of removed workpiece material (V_w) has to be considered as well:

$$m_{cf,s} = \dot{m}_{cf,s} \cdot \frac{V_w \cdot n_{ws}}{\rho_{cf}} \tag{4.72}$$

In order to reduce the amount of dragged out cutting fluid, a de-oiling process is often used in practice. The de-oiling of scrap material can be performed by different methods as presented in Sect. 2.3.4.2. Especially the use of centrifugation is common and the recovered cutting fluid can be reused without restrictions and returned to the system, but the use of such a process is also connected to additional energy consumption. For this purpose, a modeling approach proposed by Kloke (2003) for a rotating cleaning system is used and adapted. The principle structure and relationships of the de-oiling process for the model formulation are illustrated in Fig. 4.26.

Accordingly, the drive power of the vertically rotating cylinder for the de-oiling process (P_{de}) can be formulated as a function of rotational speed (n_s), acceleration time ($t_{de,a}$) and moment of inertia (J):

$$P_{de} = 4 \cdot \pi^2 \cdot n_s^2 \frac{J}{t_{de,a}} \tag{4.73}$$

The acceleration time is usually assumed to be five seconds (Riegert and Mager 2001). The rotational speed has to be determined to remove unbound cutting fluid between the chips as well as bound fluid adhered to the surface of the chips, which requires much higher speeds. It therefore depends on the adhesion force (F_{Ad}) between fluid and chip surface, centrifuge radius (R_c), limit radius (r_L) and fluid mass ($m_{cf,s}$) that needs to be removed:

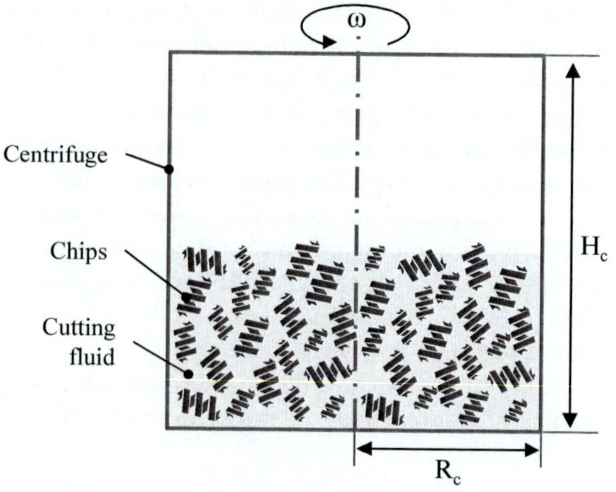

Fig. 4.26 Relationships of the de-oiling process [according to Kloke (2003)]

$$n_s = \frac{1}{2 \cdot \pi} \cdot \sqrt{\frac{F_{Ad}}{m_{cf,s}} \cdot \frac{1}{R \cdot \sqrt{1 - (1 - \frac{r_L^2}{R_c^2})}}} \tag{4.74}$$

The adhesion force further depends on the angular velocity (ω) and is described by the following equation:

$$F_{Ad} = \omega^2 \cdot m_{cf,s} \cdot r_L \tag{4.75}$$

The moment of inertia considers the chips with a specific mass (m_c) as well as the centrifuge with its material density (ρ_c), thickness of the shell (s_1) and thickness of the bottom part (s_2).

$$J = \frac{1}{2} m_c \cdot R_c^{\ 2} + 2\pi \cdot s_1 \cdot \rho_c \cdot H_c \cdot \left(R_c + \frac{s_1}{2}\right)^3 + \frac{1}{2}\pi \cdot s_2 \cdot \rho_c \cdot (R_c + s_1)^4 \tag{4.76}$$

The drivetrain of the centrifuge is not considers as its impact is comparatively low (Kloke 2003).

As presented in Table 2.5, the level of residual fluid is around 15–20% for the grinding swarf and 1–5% for turning and milling chips after using the centrifugation method. The degree of wetting before this de-oiling process, however, has no impact on the results (Riegert and Mager 2001).

Besides the scrap material, the workpieces are often treated as well after the machining process. They have to be cleaned from chips and cutting fluids in preparation for painting, phosphating or coating as well as for intermediate cleaning between different machining processes (Mang and Dresel 2007). For this purpose, the immersion cleaning is preferable and sufficient for further processing of machined parts. For this cleaning method, chemicals are used in a circulating system, but the washed off cutting fluid is lost and cannot be returned to the system. Based on a life cycle assessment, it has been demonstrated that the energy consumption of the cleaning unit e.g. for tempering the cleaning medium has the highest impact and is particularly relevant. The temperature can vary from 50 to 80 °C. Compared to this, the impact for the production of cleaning chemicals is rather low (Wilbert et al. 1998). Similarly, Schrems (2014) also focused on immersion cleaning as the relevant method to remove cutting fluid and chips from workpieces and indicated that the heating system accounts for around 70–90% of the energy demand. The heating system of the cleaning unit can be modeled using an approach that determines the heating of swimming pools (VDI 2089 2010). For both heating systems, the heating load for the initial heating of a fluid volume to a certain temperature and the heating load during operation are relevant. The initial heating load ($Q_{cm,i}$) is determined as follows:

$$Q_{cm,i} = \rho_{cm} \cdot c_{p,cm} \cdot V_{cm} \cdot \Delta t_{cm} \tag{4.77}$$

It depends on the fluid properties specific heat capacity ($c_{p,cm}$) and density (ρ_{cm}) as well as the fluid volume (V_{cm}) and difference in temperature (Δt_{cm}). Due to the high water content of the cleaning fluid, similar properties to water can be assumed. During operation, the heat load is determined by the evaporation heat loss for the evaporation of the cleaning fluid:

$$\dot{Q}_{cm,ev} \approx \dot{m}_{cm} \cdot r \approx \frac{\beta_m}{R_{D,cm} \cdot \overline{T}} \cdot \left(p_{v,cm} \cdot p_{v,air}\right) \cdot A_{cm} \cdot r \tag{4.78}$$

The evaporated mass of the cleaning fluid ($\dot{m}_{cm}$) is determined by the fluid transfer coefficient (β_m), specific gas constant for vapor ($R_{D,cm}$), arithmetic mean of fluid and air temperature ($\overline{T}$), saturation pressure of vapor at fluid temperature ($p_{v,cm}$), fluid vapor pressure of air temperature ($p_{v,cm}$), surface area of the cleaning tank (A_{cm}) and evaporation heat at fluid temperature (r). Regarding the fluid transfer coefficient, different reference values are provided by VDI 2089 (2010). Although these values and the approach refer to water, it is assumed that they can also be transferred to cleaning fluids, as described above. Winter (2016) also successfully applied this approach to determine the drag-out of cutting fluids via evaporation from the tank. Further, the heat losses due to heat radiation and convection are also considered according to Eq. 4.57–4.64 for the surface of the immersion cleaning system.

However, in the case of dry machining, when workpieces mainly need to be cleaned from chips, this cleaning method is sometimes replaced by the use of surface air cleaners for the intermediate cleaning between different machining processes (Madanchi et al. 2015b). These cleaners use compressed air to blow off the chips from the workpiece. The compressed air demand has to be converted from liters to watt hours to make it comparable with the direct electricity demand. For simplification, a compressor-specific conversion factor can be used that is independent from the pressure level. Alternatively, the power demand of the compressed air consumption can also be determined based on the physical approach used by Züst et al. (2013):

$$P_{comp} = \frac{1}{\eta_{comp}} \left[V_n \cdot \rho_n \cdot R_D \cdot t_{in} \cdot ln\left(\frac{p_{comp}}{p_{in}} \right) \right] \quad (4.79)$$

It considers the compressor efficiency (η_{comp}), demanded amount of compressed air (V_n), pressure (p_{in}) and temperature (t_{in}) to demonstrate the inlet air conditions and the desired pressure level of the compressed air (p_{comp}). Further, gas properties are represented by the specific gas constant (R_D) and the density (ρ_n).

4.2.4 Multiple Machining Systems and Process Chains

The aforementioned models of the different machining system elements can be applied to analyze a single machining system. It allows for the investigation of alternative system designs and parameter settings. In practice, however, it is often required to use multiple machining processes within a process chain to produce a defined workpiece. In this case, the presented models of the elements are also valid, but other aspects also need to be considered. As presented in Sect. 2.4, the decision for centralized or decentralized peripheral subsystems as well as the need for additional cleaning and scrap treatment processes is critical and depends on the selected cutting fluid strategy. Table 4.1 lists these subsystems and processes and evaluates to what extent they are relevant for the strategy.

In view of dry machining, the use of a cutting fluid systems is not required, whereas the exhaust air system shall always be used to ensure occupational health and safety (VDI 3802-2 2012). In contrast to flooding, it is further not required to treat the scrap material by de-oiling processes. The cleaning of the workpieces is usually needed regardless of the cutting fluid strategy to remove residual material. However, in case of dry machining, surface air cleaners using compressed air are often used alternatively to cleaning fluids. By using MFOs it is even possible to eliminate workpiece cleaning between machining processes due to the compatibility of the fluids (Madanchi et al. 2015b). Overall, these are rather logical connections and constraints than mathematical models, but still need to be integrated into the concept.

Further, it is essential to adapt the dimensioning of the peripheral subsystems to support multiple machining systems or an entire process chain. Regarding the cutting fluid system the dimension of the system tank is particularly relevant for cooling and avoiding microbial contamination. The fluid volume can be determined using the residence time (t_r), which is recommended at 6 to 10 min for water-based and 10 to 20 min for non-water-based cutting fluids as well as the recirculation number, which is recommended at 6 to 10 times per hour for water-based and 3 to 6 times per hour for non-water based cutting fluids (VDI 3035, 2008). The total cutting fluid volume ($V_{cf,total}$) is composed of the minimal tank volume (V_{tv}) and the travel volume (V_t):

$$V_{cf,total} = V_{tv} + V_t \quad (4.80)$$

Table 4.1 Requirement for peripheral subsystems and relevant processes depending on cutting fluid strategies [according to Madanchi et al. (2015b)]

Peripheral subsystems and relevant processes	***Cutting fluid strategies***		
	Flooding	Flooding (using MFO)	Dry
Cutting fluid system			
decentralized	●	●	○
centralized	●	●	○
Exhaust air system			
decentralized	●	●	●
centralized	●	●	●
Scrap treatment process	●	●	○
Cleaning (between machining processes)			
washing fluid	●	○	◐
compressed air	○	○	◐
Cleaning (between different processes)			
washing fluid	●	●	◐
compressed air	○	○	◐
Cleaning (end of process chain)			
washing fluid	●	●	◐
compressed air	○	○	◐

● applicable ◐ limited applicable ○ not applicable

The minimal tank volume is determined by the residence time and cutting fluid volume flow. For the supply chain it is assumed that all machine tools could operate simultaneously and be supplied with cutting fluid:

$$V_{tv} = t_r \cdot \sum_{mt=1}^{MT} \dot{V}_{cf,mt} \tag{4.81}$$

The travel volume considers the amount of cutting fluid within the pipes. On the assumption of a possible simultaneous operation of the machine tools, all pipes from the tank to each machine tool are considered:

$$V_t = \sum_{mt=1}^{MT} \sum_{e=1}^{E} \pi \frac{d_{i,e,mt}^2}{4} \dot{V}_{cf,mt} \tag{4.82}$$

The volume within each piping element (e) is determined based on the inner diameter (d_i). Based on this approach, the initial cutting fluid demand of the system is described, but the continious cutting fluid demand results from the drag-out by emissions and adhering to chips and workpieces.

In the case of centralized peripheral systems, it is further needed to identify the proportional demand of each machining system (Winter 2016). For the cutting fluid system, the specific power demand is determined as a ratio of the volume flow of the machine tool to the total volume flow:

$$P_{P,el,mt} = \frac{P_{P,el,total}}{\dot{V}_{cf,total}} \cdot \dot{V}_{cf,mt} \tag{4.83}$$

This is similar for the exhaust air system, but based on the ratio of the air flow ($\dot{V}_{ea}$) of the machine tool to the total air flow:

$$P_{ea,mt} = \frac{P_{ea,total}}{\dot{V}_{ea,total}} \cdot \dot{V}_{ea,mt} \tag{4.84}$$

4.2.5 Model Overview and Connection

The derived models for the different elements are combined to represent the machining system. It is considered to be a modular structure, where elements can be eliminated or added to the machining system depending on the considered strategy and scope. A variety of different input parameters are defined for the models, which can simultaneously affect different models. Moreover, the output of a system element model can also be the input for another system element model. These interactions between models and input parameters are illustrated in Fig. 4.27. This illustration demonstrates the case of a machining system using flooding as cutting fluid strategy. The machining process is considered generically and only differs with regard to surface quality, as additional process parameters are taken into account for grinding compared to turning and milling. The input parameters are differentiated in terms of workpiece and tool specific parameters, cutting process parameters, cutting fluid specific parameters and others. The workpiece and tool properties are generalized for this figure and include both the geometry and the material. The same applies to the properties of the cleaning, centrifuge and machining system geometry. In detail these are rather diverse and differ for the specific model. In contrast, for the process parameters and fluid properties the specific input parameters can be precisely identified, which affect a large number of the models. From this it can be concluded that by

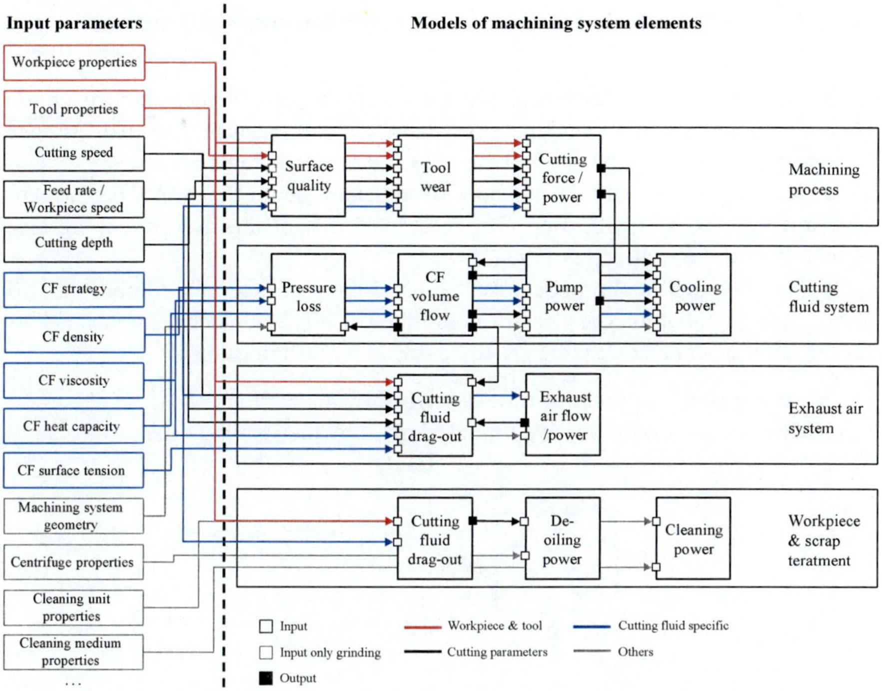

Fig. 4.27 Connections between models of machining system elements

adjusting these input parameters a great influence on the performance of the overall system is achieved.

As part of the concept, the models are to be experimentally validated to ensure sufficient accuracy and quality. This chapter demonstrates that this has already been accomplished for a variety of the demonstrated models. However, the used case study did not allow for a verification of the tool wear and subsequent subsystems of de-oiling and cleaning. Moreover, the comparison between theoretical and experimental results showed some larger deviation for the grinding power and surface roughness during milling. For this reason it is recommended to further extend the experiments and possibly include further input parameters into the model. It is also pointed out that the drag-out is determined by empirical regression models, which may require additional effort to be transferred to alternative machining systems. Further adjustments could also be made in this context.

4.3 Environmental, Economic and Technological Impact

As presented in Sect. 2.5.2, it is required to evaluate the performance of the machining system from an environmental, economic and technological perspective. By using the presented modeling approach, the energy and resource flows can already be determined for each system element and need to be quantified for the overall performance assessment in the following step. The relevant flows for this assessment are again summarized and visualized in Fig. 4.28.

In order to examine the impact for the production of a certain number of workpieces, it is essential to consider different time periods and machine states. For this concept, it is assumed that there are three relevant machine states:

- processing state - productive time described by the cutting time (t_c)
- standby state - non-productive time described by the standby time (t_s)

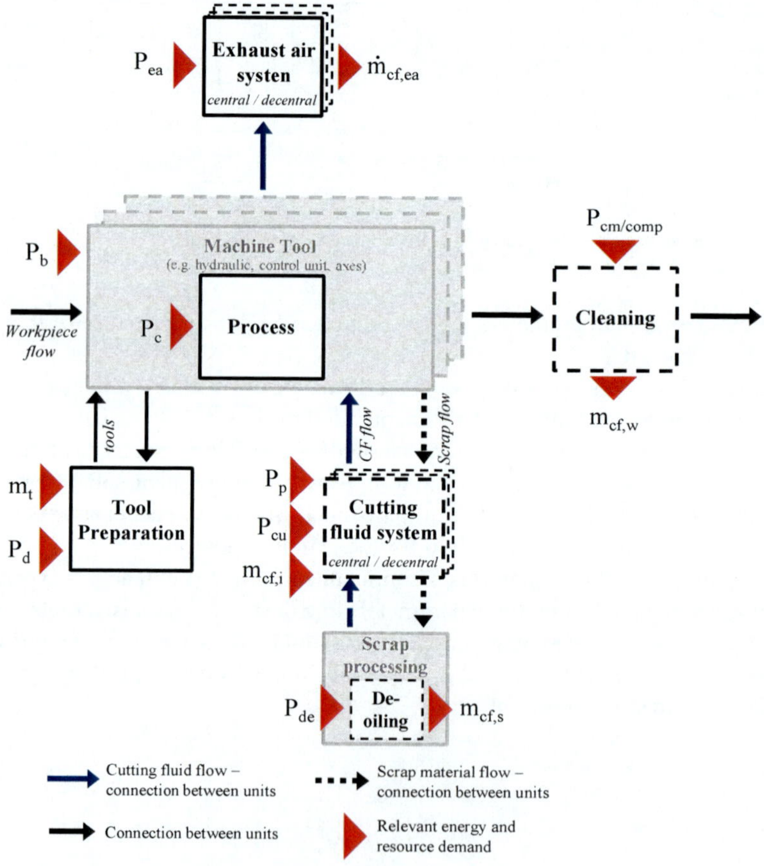

Fig. 4.28 Energy and resource flows of the relevant subsystems

- preparation state - non-productive time described by the machine downtime (t_d) due to dressing, tool or workpiece changes.

As also visualized in the energy break downs (see Fig. 4.5), all subsystems are active for the processing state and the actual material removal occurs. In the standby state the machine tool is operational, but non-productive due to e.g. waiting. At this state, the machine tool runs at base load without an active spindle and other subsystems. Usually, the exhaust air system is active at this state, while the cutting fluid system recirculates but is not used to supply fluid to the machine tool. Similarly, the machine tool runs at base load for the preparation state as well. In line with the approach by Pusavec et al. (2010), this results in the following production time (t_p):

$$t_p = t_c + t_s + t_d \tag{4.85}$$

The cutting time is determined based on the material volume to be removed from the workpiece (V_w) and the material removal rate (Q_w) defined by the process parameters:

$$t_c = \frac{V_w}{Q_w} \tag{4.86}$$

The standby time of the production depends on the specific use case. The machine downtime comprises the time for tool preparation (t_{tp}) and workpiece preparation (t_{wp}):

$$t_d = t_{tp} + t_{wp} \tag{4.87}$$

Regarding machining processes with geometrically defined cutting edges the time for tool preparation describes the time to change tools and inserts. In case of turning, an average tool change time of 2 min was measured by Mativenga and Rajemi (2011) and also adapted from Priarone et al. (2016). In the case of grinding, the tool preparation time focuses on the time required for dressing and conditioning the grinding wheel. According to Winter (2016), it is determined based on the tool width (b_s), traverse dressing speed (v_{fad}), radial grinding wheel wear (Δr_s) and selected dressing depth (a_{ed}):

$$t_{tp,grinding} = \frac{b_s}{v_{fad}} \cdot \frac{\Delta r_s}{a_{ed}} \tag{4.88}$$

Besides the production time it is further required to take the de-oiling time (t_{de}) and the cleaning time of the workpiece (t_{cm}) into account as well.

Environmental impact

Within this concept, the focus is on the use stage of machining systems. Similar to other research studies assessing the environmental impact of machining (Herrmann et al. 2017; Winter 2016), the reference flow considers a certain number of produced workpieces over a certain time. The exact values, however, depend on the specific use case. In general, the assessment follows a simplified LCA-approach. For the impact assessment, a number of different categories such as water use, aquatic toxicity, acidification or GWP can be applied. The conducted literature review, however, demonstrated the examination of GWP is common to most studies focusing on machining systems in connection with cutting fluid strategies. It is also known as carbon footprint and describes the gradual warming of the atmosphere, which as a consequence leads to climate change. Carbon dioxide serves as a reference substance and the unit of relevant flows has to be converted into equivalents (kg CO_2-eq) (Herrmann 2010). In this case, the kg CO_2-equivalents for electrical energy ($CO_{2,e}$) is used for the direct energy consumption of the subsystems (P_{ea}, P_p, P_{cu}, P_c, P_b, P_d, P_{cm}, P_{de}). This value depends on the specific energy mix that is being used. While the kg CO_2-equivalent for cutting fluid production ($CO_{2,cf,p}$) is used for the fluid drag-out and relatively for the tank volume ($m_{cf,i}$, $m_{cf,w}$, $m_{cf,s}$, $m_{cf,ea}$), the kg CO_2-equivalent for cutting fluid disposal ($CO_{2,cf,d}$) is only used relatively for the tank volume ($m_{cf,i}$). The kg CO_2-equivalent for tools ($CO_{2,t}$) is used for the used tools (m_t). The values for the CO_2-equivalent can be taken from literature or databases such as Ecoinvent and CES material selector.

The overall environmental impact of a machining system (EI_{ms}) is therefore composed of the respective impacts resulting from energy (EI_e), cutting fluid (EI_{cf}) and tools (E_t) as described by the following equation:

$$EI_{MS} = EI_{cf} + EI_t + EI_e \tag{4.89}$$

The environmental impact due to cutting fluid is determined as follows:

$$\begin{aligned} EI_{cf} = &\left(\left(m_{cf,i} \cdot \frac{t_p}{t_{cf,1}}\right) + m_{cf,w} + m_{cf,s} + \left(\dot{m}_{cf,ea} \cdot t_c\right)\right) \\ &CO_{2,cf,p} + \left(m_{cf,i} \cdot \frac{t_p}{t_{cf},l}\right) \cdot CO_{2,cf,d} \end{aligned} \tag{4.90}$$

The different possibilities of cutting fluid drag-out are determined according to the models, which are described in the previous section. The impact of the initial cutting fluid filling ($m_{cf,i}$) is only partially applied. In this case, the service life of the cutting fluid ($t_{cf,l}$) is to be taken into account as well as the considered production time. With the use of comprehensive care measures, the service life of mineral oil based emulsion is around 20 weeks (Kiechle 1996). However, when non-water-based cutting fluids are used, the service life even exceeds a year. Following the LCA-approach, the CO_2-equivalent for the cutting fluid comprises the raw materials, production and disposal of the cutting fluid (Dettmer 2006). This results into different values depending on the type of cutting fluid. For dry machining this term does not apply.

The environmental impact resulting from the tools is calculated based on the following equation:

$$EI_{t,i} = m_t \cdot CO_{2,t}$$
$$\text{For turning and milling}: m_t = \frac{t_c}{t_l} \tag{4.91}$$
$$\text{For grinding}: m_t = \frac{V_{gt}}{V_{gw}}$$

For turning and milling processes, the tool use (m_t) is determined based on the ratio of the possible tool life (t_l) to the cutting time (t_c). Here, only this actual cutting time has an effect on the tool wear. For grinding, the tool use is calculated based on the previously modeled grinding wheel wear volume (V_{gt}) related to the total grinding wheel volume (V_{gw}). The CO_2-equivalent for the tools again takes the used raw materials, production and disposal into account and depends on the specific tool and material.

The impact of energy use is determined taking into account the energy demand of the different machine states and processes:

$$EI_e = (E_c + E_s + E_{t,i} + E_{cu} + E_{de} + E_{cm}) \cdot CO_{2,e} \tag{4.92}$$

For different machining systems, this energy demand was already visualized in the previously presented energy break downs at the standby and processing state. It also presented the base load of the machine tool (P_b), which depends on the specific machine tool and is usually not influenced by the selected process parameters and cutting fluid strategy. It is therefore considered as a constant value.

For the processing state, the energy demand is determined as follows:

$$E_c = \left(P_{ea} + P_{p,p} + P_c + P_b\right) \cdot t_c \tag{4.93}$$

Regarding the standby state it is generally assumed that the machine tool runs at base load, but with an active exhaust air system. In some cases the exhaust air system needs a few minutes to start up again after it has been switched off. Regarding the cutting fluid system, it might be possible that a pump is still used to filter or to circulate the cutting fluid within the system:

$$E_s = \left(P_{ea} + P_{p,s} + P_b\right) \cdot t_s \tag{4.94}$$

At preparation state the energy demand for turning and milling is determined similarly.

$$E_{t,i} = \left(P_{ea} + P_{p,s} + P_b\right) \cdot t_d \tag{4.95}$$

In the case of grinding, however, the dressing process has to be taken into account as well. For this process the peripheral subsystems are required and a dressing power (P_d) is applied. For simplification, this is assumed as a given value.

$$E_{t,g} = \left((P_{ea} + P_p + P_b + P_d) \cdot t_{tp}\right) + \left((P_{ea} + P_{p,s} + P_b) \cdot t_{wp}\right) \tag{4.96}$$

The cooling unit is simply determined by the required power of the cooling unit (P_{cu}) over the production time:

$$E_{cu} = P_{cu} \cdot t_p \tag{4.97}$$

The energy use for the de-oiling process considers the scrap material for the workpiece in relation to the capacity of the centrifuge and is simply determined as follows:

$$E_{de} = P_{de} \cdot t_{de} \cdot \frac{m_c}{m_{de}} \tag{4.98}$$

The energy demand for cleaning is determined in a similar way, by considering the initial heating load and possible evaporation heat loss during operation and especially during workpiece removal. However, for dry machining the power demand of compressed air (P_{comp}) might be used instead when surface air cleaners are used.

$$E_{cm} = \left(Q_{cm,i} \cdot \frac{m_w}{m_{de}} \cdot \frac{t_{cm}}{t_{cm,total}}\right) + \left(\dot{Q}_{cm,ev} \cdot \frac{m_w}{m_{de}} \cdot t_{cm}\right) + \left(\dot{Q}_{cm,ev,out} \cdot t_{cm,out}\right) \tag{4.99}$$

Economic impact

The use of energy and materials within the machining system is also connected to an economic impact. As presented in Sect. 3.3, this perspective is not considered as widely in research approaches as the other perspectives, although it is particularly relevant in practice. It is usually the decisive perspective that decides on investments for improvement measures, which are evaluated based on the possible savings and return on investment. In order to support decision-making, it is therefore important to assess the economic impact comprehensively.

The method of material flow cost accounting presented in Sect. 2.5.2 involves the monetary quantification of inputs and outputs of processes and systems. It is designed for the requirements of production systems with a focus on the use phase and is used for similar applications by Madanchi et al. (2019b) and Winter (2016). As this concept analyzed the interactions of a machining system by the input and output flows, this approach also qualifies for it. The energy and material flows presented in Fig. 4.28 are again the basis for this evaluation and a specific price unit has to be assigned to them. In this case, the price for electrical energy (p_e), the purchasing price ($p_{cf,p}$) and disposal price ($p_{cf,d}$) for cutting fluids, the price for tools (p_t) as well as machine usage rate (p_m) and labor rate (p_l) are relevant. The specific values are use case specific and can be taken from internal data, literature or databases.

In line with the environmental impact, the overall economic impact of a machining system (CI_{ms}) also consists of the costs resulting from energy (CI_e), cutting fluid

(CI_{cf}) and tools (CI_t). Further, this perspective also considers the impact from machine tool usage and labor costs (CI_{ml}):

$$CI_{ms} = CI_{cf} + CI_t + CI_e + CI_{ml} \tag{4.100}$$

The economic impact due to cutting fluid use is calculated similar to the environmental impact. However, in this case, the costs for the cutting fluid disposal are also relevant.

$$CI_{cf} = (((m_{cf,i} \cdot \frac{t_p}{t_{cf,l}}) + m_{cf,w} + m_{cf,s} + (\dot{m}_{cf,ea} \cdot t_c)) \cdot p_{cf,p}) + ((m_{cf,i} \cdot \frac{t_p}{t_{cf,l}}) \cdot p_{cf,d}) \tag{4.101}$$

Moreover, the general approach for the economic impact due to tool use is the same:

$$CI_{t,i} = m_t \cdot p_t \tag{4.102}$$

This also applies for the impact resulting from energy use:

$$CI_e = (E_c + E_{st} + E_{t,i} + E_{de} + E_{cm}) \cdot p_e \tag{4.103}$$

The machine tool usage and labor costs are particularly relevant for the economic impact and rather individual for the specific machining system. Following the approach by Pusavec et al. (2010), the machine tool usage cost rate ($p_{m,i}$) comprises the investments for machine tool and peripheral subsystems, but also costs for e.g. maintenance or insurance costs. The labor cost rate includes indirect labor, fringe benefits and supervision costs besides the direct labor costs ($p_{l,i}$). Overall, the costs are determined for the production, de-oiling and cleaning processes as follows:

$$CI_{ml} = \left(t_p \cdot \left(p_{m,p} + p_{l,p}\right)\right) + \left(t_{de} \cdot \left(p_{m,de} + p_{l,de}\right)\right) + \left(t_{cm} \cdot \left(p_{m,cm} + p_{l,cm}\right)\right) \tag{4.104}$$

For each process, the cost rates are determined individually and applied over the time.

Technological impact

Machining processes are conducted to achieve defined changes on a workpiece and therefore to achieve a technological impact. As presented in Sect. 2.5.2, it can be distinguished between a process-specific and a product-specific perspective. Both perspectives can be applied within the developed concept, but the choice usually depends on the specific use case. Process-specific indicators such as cutting power or cutting time are modeled in the concept and are more relevant for turning and milling. These processes are usually followed by finishing processes, so the focus is on the process rather than the product. In contrast, grinding is considered as such a finishing process and therefore focuses more on the product and the quality of the output. As presented, there are different indicators to express this workpiece quality, which depend on various influencing parameters and are therefore difficult to predict. For this concept, the surface roughness of the workpiece is used as product-specific indicator, as its relation to the process parameters is generally examined and modeled. The models to determine the arithmetic mean roughness value are presented in Sect. 4.2.2.2 for turning, milling and grinding and can therefore be used to express the product-specific technological impact.

Indicators are essential to support decisions between different process or design alternatives or to evaluate the potential of improvement measures. The selected indicators to assess the performance of the machining systems are again summarized in Fig. 4.29.

These indicators can be applied as an independent single-objective or multi-objective approach. By nature, the three perspectives are mostly in conflict with each other, making it difficult to achieve an optimal solution. In case of a multi-objective approach it is possible to consider all perspectives and to find a compromise between them. For this purpose different methods and approach are available. Kuram et al. (2013) used one indicator as the primary objective function and the other indicators as implicit constraints. The approach presented by Wang et al. (2014) aimed at minimizing the indicators simultaneously, while using the cutting parameters constraints, so that they can only be selected within a certain range. A similar approach was also used by Jin et al. (2009), but in this case the importance of each indicator was incorporated by a weight ranging from 0–1.

In view of this concept, the indicators are aggregated into an overall composite index. By weighting the different indicators and their priority it is possible to combine

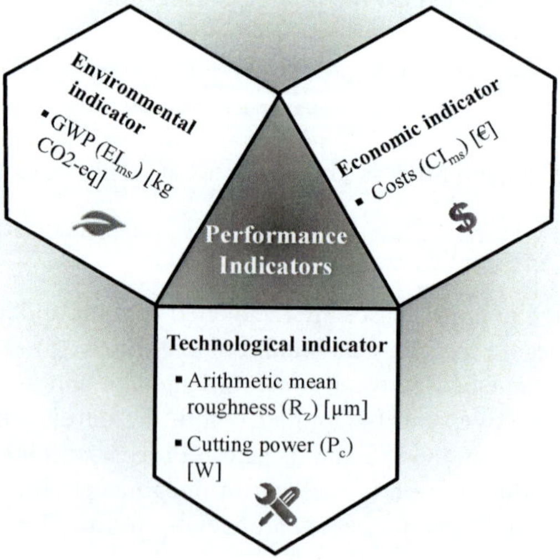

Fig. 4.29 Performance indicators to assess machining systems

them into a single linear objective function, which can be optimized against reference values. Various constraints regarding the range of possible cutting parameters or cycle times are also taken into account for the optimization.

4.4 Improvement Measures

Following the performance assessment of the machining system from different perspectives, the fourth module of the concept aims to improve it (see Fig. 4.1). For this purpose, different measures are available, which differ with regard to their effectiveness and effort. In general, these improvement measures can be classified as

- organizational measures with a focus on process planning and
- technical measures with a focus on hardware and components.

With regard to the energy demand, Zein et al. (2011) presented the impact of both strategies on energy reduction using Fig. 4.30.

Accordingly, technical measures lead to a lower power demand of the systems by e.g. replacing components with more efficient ones. The organizational measures, however, lead to a reduction in the time of use of systems by e.g. selecting more suitable process parameters. These correlations can also be transferred to resources. Further, improvement measures can be distinguished by their focus level. They can aim at the process or different peripheral subsystems, but also simultaneously at multiple subsystems.

The considered measures also need to be classified according to

- implementation effort of the measure and
- improvement potential regarding all three perspectives.

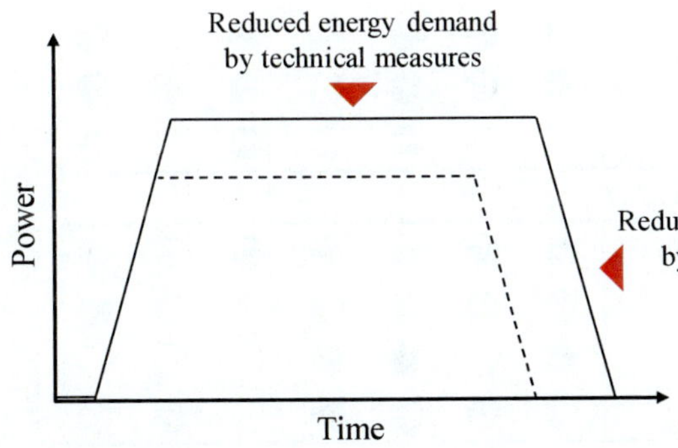

Fig. 4.30 Impact of organizational and technical measures on energy demand [Zein et al. (2011) based on Fanuc (2008)]

According to these improvement strategies and classification, selected measures are shown in Table 4.2 and described in the following.

4.4.1 Description of Measures

The organizational measure to improve machining systems by selecting more suitable process parameters is widely used. Following the experience-guided approach for machining process with defined cutting edges presented in VDI 3321(1994), it is recommended to first adapt the cutting depth, then the feed rate and finally the cutting speed due to the highest impact on tool wear. The literature review also demonstrated that there are various approaches that adapt these parameters by using a multi-objective optimization approach considering different performance indicators

Table 4.2 Overview of possible improvement measures

Improvement measures	Process parameters	Standby strategy	Cutting fluid type	Nozzle design	Efficient components	Emission reduction	De-/centralized design
Improvement strategy							
Technical approach	○	○	●	●	●	●	●
Organizational approach	●	●	○	○	○	○	○
Improving object							
Machining process	●	○	●	●	○	○	○
Cutting fluid system	●	●	●	●	●	○	●
Exhaust air system	●	●	●	○	●	●	●
WP & scrap system	○	◑	●	○	●	○	●
Improving potential and effort							
High environmental potential	●	●	●	●	●	◑	●
High economic potential	●	●	●	◑	◑	◑	●
High technical potential	●	○	●	●	○	○	○
Low implementation effort	◑	●	◑	●	○	●	○

● applicable ◑ slightly applicable ○ not applicable

such as energy, surface quality, tool life or costs (Kuram et al. 2013; Li et al. 2017; Wang et al. 2014; Yan and Li 2013). For this concept, the objective is basically formulated by the following model:

$$Z = \mathrm{Min}(EI_{ms}, CI_{ms}, R_a)$$

Subject to

$$\begin{aligned} v_{c,\min} &\le v_c \le v_{c,\max} \\ f_{\min} &\le f \le f_{\max} \\ a_{e,\min} &\le a_e \le a_{e,\max} \end{aligned} \tag{4.105}$$

It aims at minimizing the impact of the environmental, economic and technological indicators by adapting the cutting parameters within a certain range. Instead of surface roughness, other technological indicators such as cutting power can also be considered. Further constraints can be added or changed depending on the specific use case. Applying this measure has an impact on all three perspectives and the entire machining system but not on the following de-oiling and cleaning processes.

The organizational approach to improve the performance by switching off peripheral subsystems that are not required in the standby and preparation state is also discussed in literature for certain scenarios. Eisele (2014), for example, analyzed this measure for a hydraulic system and revealed energy savings of around 40% for a specific use case. It is important that the times to ramp up and down the component are taken into account. The measure aims to impact the environmental and economic but not the technological perspective and can be implemented with little effort in the control unit.

With regard to the aim of this concept, adapting the cutting fluid strategy has a high impact on each subsystem and assessment perspective. The research approach by Winter et al. (2015) demonstrated that the grinding power is about 115 W/mm lower for a mineral oil based emulsion compared to a grinding oil and thus corresponds to a relative energy saving potential of about 48% as well as an absolute saving of 1.15 kW for a specific use case. The previous literature review also presented various other research approaches that evaluated the impact of different strategies for different perspectives.

Another technical measure focuses on adapting the design of cutting fluid nozzles. It is particularly relevant for grinding processes and is intended to reduce the required pump power and to improve the volume flow in the contact zone. A study by Madanchi et al. (2017) demonstrated that under the assumption of a required volume flow in the contact zone, a point jet nozzle saves, for example, 23.5% of the pump power compared to a slot nozzle, resulting in an absolute energy saving of less than 0.1 kW.

The replacement of components by more efficient ones is another technical measure. As demonstrated in Sect. 2.3.2, the use of a high efficiency motor (classification from IE1 to IE4) can increase the efficiency. The use of a 2 kW motor with the energy efficiency class IE4 instead of IE1 can, for example, increases the efficiency by around 13% and thus achieves an absolute saving of around 0.4 kW (DIN EN 60034-30-1 2014). This measure therefore aims at environmental and economic savings without influencing the technological results. Especially for improving existing machining systems this measure is also associated with higher investment costs and thus greater implementation effort.

An improvement measure focusing on the exhaust air system deals with the reduction of cutting fluid emissions. Madanchi et al. (2019a) demonstrated that the drag-out of cutting fluid for a specific grinding process is around 75.2 g/h and that there is a large share of particles with a size of 2 μm and higher, which are particularly critical in terms of occupational health. These emissions could be reduced by using a bell-shaped film of cutting fluid around the process to capture the particles. This only has a moderate environmental and economic impact, but requires only little implementation effort.

The measure to use centralized or decentralized peripheral subsystems is particularly relevant for the planning of new machining systems within a process chain as it is connected with high implementation effort. It is not intended to have an impact on the technological results, but to realize environmental and economic savings. Madanchi et al. (2015b), for example, demonstrated that for the considered process chain the use of centralized peripheral subsystems results into 12 kW less electrical energy and thus 5% savings compared to a centralized system.

Overall, the measures aim at improving the performance of machining systems and based on specific applications in literature, they have proven to be effective with high potential for this concept. While the saving potential is presented qualitatively in Table 4.2, the actual quantitative impact of specific improvement measures is presented in the next chapter for different use cases.

4.4.2 Impact Visualization

In order to visualize the impact of changed parameter settings and alternative strategies, chord diagrams are used as part of this concept. This type of diagram is a graphical method to display flows and relationships between different entities. The schematic structure for application within the concept is demonstrated in Fig. 4.31. The diagram is divided into two parts. The input parameters that are in focus are listed below. It has been revealed that especially the process parameters of cutting depth, feed rate and cutting speed as well as cutting fluid strategy are relevant for the models of the concept (see Fig. 4.27). The cutting fluid strategies also include different fluid properties. The upper part of the diagram demonstrates the resulting energy and resource flows from the machining system, which are also summarized in Fig. 4.28 as well as the production time and the overall performance indicators for the three considered perspectives. Connecting flows between the lower and the upper part demonstrate the relationship between input parameter and result while the width of the flow indicates the impact intensity. For the determination of this impact, the resulting values of the basic system settings are used as reference values.

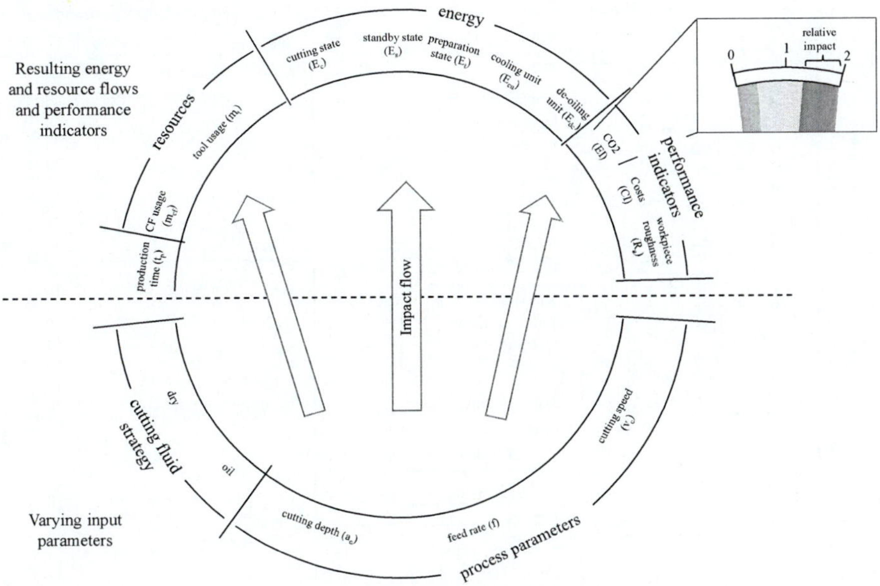

Fig. 4.31 Schematic structure of impact visualization

By changing an input parameter, a result value may also change, whereby the change is determined in relation to the reference value. Therefore doubling the result value means a relative impact of 1. The relative impact value is not limited and the greater the value, the greater the impact.

The application of this diagram is shown in the next chapter for the case studies. It is an intuitive method to visualize the impact and relevance of parameter changes on different results.

4.5 Concept Implementation

The described concept for energy and resource efficient machining systems is intended to be applied by various stakeholders for different purposes. It is therefore designed modularly with the possibility for extensions or simplifications, but always aiming to enhance the system understanding of planned and existing machining systems as well as to reveal improvement potentials for it. To guide users through the process of applying the concept and models, an application procedure is described in Fig. 4.32. In order to further support the application, the implementation into an integrated software tool is also demonstrated. This tool is supposed to assist users with the analysis, by automatically executing the models and facilitating data input in a structured manner. Moreover, it determines the performance of the system based

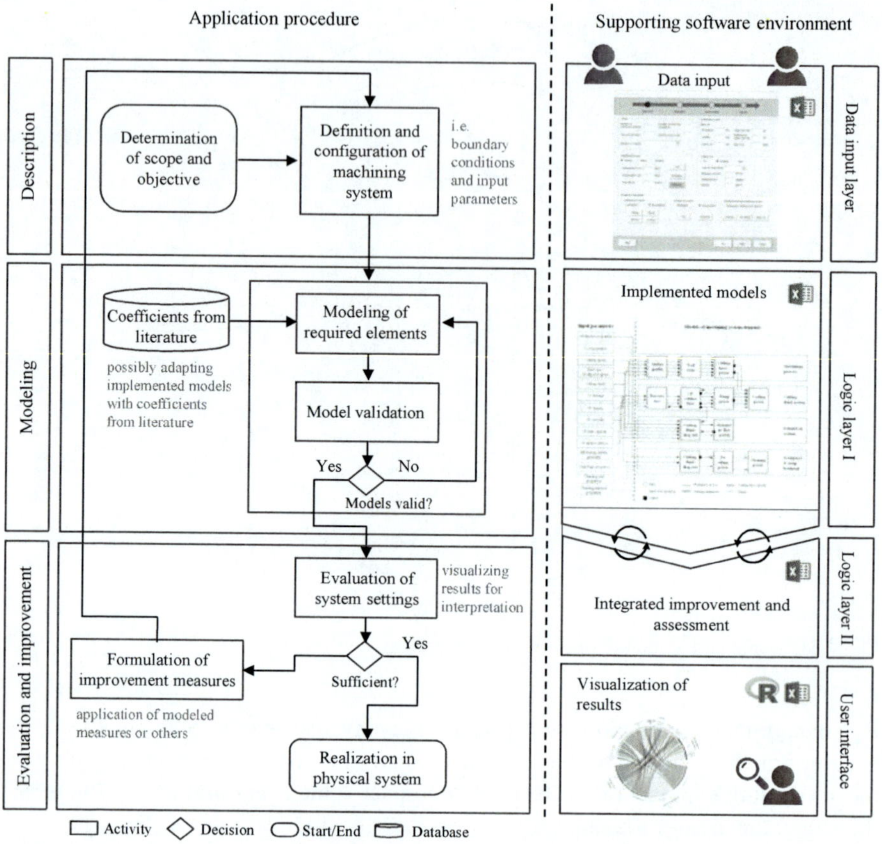

Fig. 4.32 Framework of concept implementation

on the defined indicators for each perspective and evaluates the effect of measures on it.

The general application procedure is based on the framework of the developed concept presented in Fig. 4.1. It consist of the main steps of description, modeling, evaluation and improvement. These steps are represented in the software tool by three main layers. The data input layer required for collecting, harmonizing and preparing of data, the logic layers to transfer the input data into desired performance indicators and the data output layer to visualize the results. The input and output layers are designed as graphical user interfaces to facilitate communication and a structured application. In general, these steps and layers are based on each other and follow a specific procedure.

Description & data input layer

As a first step, it is required to specify the scope and objectives of the analysis

In this regard, the planning of a single machining system or of a process chain with multiple machining systems is to be determined. Further, a general distinction is made between the analysis of a single system and the comparison of alternative settings and improvement measures to a reference system. For the purpose of performance evaluation, this also required the definition of a reference unit. Additionally, the system boundaries need to be specified whether the analysis includes for example subsequent subsystems such as cleaning and de-oiling or alternative cutting fluid strategies. On the basis of this decision, each system is described by process-specific data regarding e.g. process parameters and tools on the one hand and production-specific data regarding e.g. production profile on the other. Moreover, data input regarding the peripheral subsystems and the cutting fluid strategy are also required. Specific values to describe for example workpiece, tool and cutting fluid properties can be gathered from suppliers, product information, literature or own experience.

In order to support the collection and preparation of relevant data, an input layer is developed in MS Excel® using Visual Basic for Applications (VBA). It ensures a structured procedure, where the input data are stored for each machining system in MS Excel® and the procedure can be repeated for multiple systems.

Modeling and logic layer I

Based on the configuration and description of the relevant system elements, the modeling step is performed to determine the resulting energy and resource flows as well as technological results. The presented models from Sect. 4.2 can be applied to transfer the input data into the energy and material flows of the relevant subsystems. These models are implemented within the first logic layer of the supporting software tool. As generally applicable models are used, many empirical constants to determine for example the tool life or cutting power are automatically taken from tables from the literature based on tool and workpiece material. This underlines the low application effort. Depending on the scope and choice of cutting fluid strategy, specific models and subsystems may not be required and are therefore neglected in order to minimize modeling effort. However, depending on the scope and objective it might also be required to further integrate additional system elements and therefore also models. This is also intended and possible due to the modular structure. For this purpose, the presented modeling procedure and requirements should be considered. This means that the models have to be validated on the basis of experimental data to ensure model quality and a sufficient level of accuracy. In case the desired accuracy is not achieved, the models need to be adapted.

Evaluation and improvement & logic layer II and data output layer

On the basis of the modeled energy and resource flows as well as technological results it is possible to determine the desired performance indicators from Sect. 4.3. This determination is executed by the second logic layer of the software tool and the user interface presents the results of the analysis to the user. This comprises the presentation of performance indicators for a single system or for multiple scenarios. It is intended to visualize the results intuitively in order to support decision making and to derive further improvement measures if necessary. This also includes the impact visualization of varying process parameters on the results by chord diagrams (see Fig. 4.31) using the software environment R®.

Based on the results presented for a specific system setting, the user has to decide whether alternative settings and improvement measures are required. Overall, it is irrelevant whether the user aims at planning a new or existing machining system. The second logic layer of the software tool integrates different organizational and technical measures as presented in Sect. 4.4. The application of this logic layer depends on the scope and objective of the analysis. In this case, a reference scenario has to be determined to demonstrate the effect of alternative settings and measures in comparison to it. These measures are generally defined by the user and may comprise a single measure or also a combination of multiple measures. The adapted set of data is then again used to execute the models of the first logic layer to determine the alternative flows and values. The results of the different scenarios are stored in MS Excel®. This comprises the presentation of performance indicators for a single system or for multiple scenarios. In case additional measures are identified, the changed parameters or settings are again entered via the data input layer and the aforementioned procedure is repeated.

References

Archard JF (1961) Single contacts and multiple encounters. J Appl Phys 32:1420–1425. https://doi.org/10.1063/1.1728372

Astakhov VP, Godlevskiy V (2012) Delivery of metalworking fluids in the machining zone. In: Astakhov VP, Joksch S (eds) Metalworking fluids for cutting and grinding: fundamentals and recent advances. Woodhead Publishing, Cambridge, UK, Philadephia, USA

Boothroyd G, Knight WA (2006) Fundamentals of machining and machine tools, 3rd edn. Mechanical engineering, vol 69. CRC Taylor & Francis, Boca Raton, USA

Brücher T (1996) Kühlschmierung beim Schleifen keramischer Werkstoffe. Dr.- Ing. Dissertation, TU Berlin. IPK, Berlin, Germany

Byers JP (2006) Metalworking fluids, 2nd edn. CRC Press, Hoboken, USA

Dahmus JB, Gutowski TG (2004) An environmental analysis of machining. Proc ASME Int Mech Eng Congress and r&d Expos 643–652. https://doi.org/10.1115/IMECE2004-62600

Degner W, Lutze H, Smejkal E (2015) Spanende formung: theorie, Berechnung, Richtwerte, 17th edn. Hanser, München, Germany

Denkena B, Tönshoff HK (2011) Spanen. Springer, Berlin, Heidelberg, Germany

Denkena B, Helmecke P, Hülsemeyer L (2014) Energy efficient machining with optimized coolant Lubrication flow rates. In: 5th Machining innovations conference, vol 24, pp 25–31. https://doi.org/10.1016/j.procir.2014.07.140

Dettmer T (2006) Nichtwassermischbare Kühlschmierstoffe auf Basis nachwachsender Rohstoffe. Dr.-Ing. Dissertation, Technische Universität Braunschweig. Vulkan Verlag, Essen, Germany

Dietrich J (2016) Praxis der Zerspantechnik. Springer, Wiesbaden, Germany

DIN CLC, TS 60034-31 (2011) Drehende elektrische Maschinen - Auswahl von Energiesparmotoren einschließlich Drehzahlstellantrieben. Beuth Verlag GmbH, Berlin, Germany, DIN

DIN EN 14511-1 (2019) Luftkonditionierer, Flüssigkeitskühlsätze und Wärmepumpen für die Raumbeheizung und -kühlung und Prozess-Kühler mit elektrisch angetriebenen Verdichtern. DIN, Beuth Verlag GmbH, Berlin, Germany

En DIN, 16798-3 (2017) Energetische Bewertung von Gebäuden - Lüftung von Gebäuden. Beuth Verlag GmbH, Berlin, Germany, DIN

En DIN, 60034-30-1 (2014) Drehende elektrische Maschinen - Wirkungsgrad-Klassifizierung von netzgespeisten Drehstrommotoren. Beuth Verlag GmbH, Berlin, Germany, DIN

En DIN, ISO 4287 (2010) Geometrische Produktspezifikation (GPS) – Oberflächenbeschaffenheit: Tastschnittverfahren - Benennungen. DIN, Beuth Verlag GmbH, Berlin, Germany, Definitionen und Kenngrößen der Oberflächenbeschaffenheit

Eisele C (2014) Simulationsgestützte Optimierung des elektrischen Energiebedarfs spanender Werkzeugmaschinen. Dr.-Ing. Dissertation, TU Darmstadt. Shaker Verlag, Aachen, Germany

Fanuc GE (2008) The Environmental and Economic Advantages of Energy-Efficient Motors

Guo C, Malkin S (2000) Energy partition and cooling during grinding. J Manuf Process 2:151–157

Hecker RL, Liang SY (2003) Predictive modeling of surface roughness in grinding. Int J Mach Tools Manuf 43:755–761. https://doi.org/10.1016/S0890-6955(03)00055-5

Herrmann C (2010) Ganzheitliches Life Cycle Management - Nachhaltigkeit und Lebenszyklusorientierung in Unternehmen. Springer Verlag, Berlin, Germany, VDI-Buch

Herrmann C, Thiede S, Zein A, Ihlenfeldt S, Blau P (2009) Energy efficiency of machine tools: extending the perspective. In: 42th CIRP conference on manufacturing systems, Grenoble, France

Herrmann C, Madanchi N, Winter M, Öhlschläger G, Greßmann A, Zettl E, Schwengers K, Lange U (2017) Ökologische und ökonomische Bewertung des Ressourcenaufwands - Wassermischbare Kühlschmierstoffe. Technical Report, VDI Zentrum Ressourceneffizienz GmbH (VDI ZRE)

Jin K, Zhang HC, Balasubramaniam P, Nage S (2009) A multiple objective optimization model for environmental benign process planning. In: 16th international conference on industrial engineering and engineering management, pp 869–873. https://doi.org/10.1109/ICIEEM.2009.5344194

Johansson D, Hägglund S, Bushlya V, Ståhl J-E (2017) Assessment of commonly used tool life models in metal cutting. In: 27th international conference on flexible automation and intelligent manufacturing, Modena, Italy, vol 11, pp 602–609. https://doi.org/10.1016/j.promfg.2017.07.154

Khare SK, Agarwal S (2015) Predictive modeling of surface roughness in grinding. In: 15th CIRP conference on modelling of machining operations, vol 31, pp 375–380. Karlsruhe, Germany. https://doi.org/10.1016/j.procir.2015.04.092

Kiechle A (1996) Kostenanalyse beim Einsatz von Kühlschmierstoffen. Mineralöltechnik No. 10, Stuttgart, Germany

Kienzle O (1952) Die Bestimmung von Kräften und Leistungen an spanenden Werkzeugen und Werkzeugmaschinen. VDI-Z:299–305

Kleppmann W (2011) Versuchsplanung: Produkte und Prozesse optimieren, 7th edn. Praxisreihe Qualitätswissen. Hanser, München, Germany

Klocke F (2011) Manufacturing processes 1: Cutting. Springer Verlag, Berlin, Heidelberg, Germany

Kloke U (2003) Auslegung von Bauteilreinigungsanlagen mit Hilfe eines Fachinformationssystems. Dr.-Ing. Dissertation, TU Dortmund. Maschinenelemente Verlag, Dortmund, Germany

Knobloch H (1979) Kühlschmierstoffpflege in der Praxis. Lexika Verlag, Grafenau, Württ, Germany

König W, Essel K (1982) Spezifische Schnittkraftwerte für die Zerspanung metallischer Werkstoffe. Verlag Stahleisen, Düsseldorf, Germany

Kronenberg M (1969) Grundzüge der Zerspanungslehre. Springer Verlag, Berlin, Heidelberg, Germany

Kuhrke B (2011) Methode zur Energie- und Medienbedarfsbewertung spanender Werkzeugmaschinen. Dr.-Ing. Dissertation, TU Darmstadt. epubli, Berlin, Germany

Kuram E, Ozcelik B, Bayramoglu M, Demirbas E, Simsek E (2013) Optimization of cutting fluids and cutting parameters during end milling by using D-optimal design of experiments. J Clean Prod 42:159–166. https://doi.org/10.1016/j.jclepro.2012.11.003

Li C, Chen X, Tang Y, Li L (2017) Selection of optimum parameters in multi-pass face milling for maximum energy efficiency and minimum production cost. J Clean Prod 140:1805–1818. https://doi.org/10.1016/j.jclepro.2016.07.086

Madanchi N, Winter M, Herrmann C (2015a) Cutting fluid drag-out and exhaust air in grinding processes: influence on the eco-efficiency. In: 22th CIRP international conference on life cycle engineering, vol 29, pp 329–334, Sydney, Australia. https://doi.org/10.1016/j.procir.2015.02.054

Madanchi N, Kurle D, Winter M, Thiede S, Herrmann C (2015b) Energy efficient process chain: the impact of cutting fluid strategies. In: 22th CIRP international conference on life cycle engineering, vol 29, pp 360–365, Sydney, Australia. https://doi.org/10.1016/j.procir.2015.02.056

Madanchi N, Winter M, Thiede S, Herrmann C (2017) Energy efficient cutting fluid supply: the impact of nozzle design. In: 24th CIRP international conference on life cycle engineering, vol 61, pp 564–569, Kamakura, Japan. https://doi.org/10.1016/j.procir.2016.11.192

Madanchi N, Leiden A, Winter M, Asbach C, Lindermann J, Herrmann C, Thiede S (2019a) Cutting fluid emissions in grinding processes: influence of process parameters on particle size and mass concentration. Int J Adv Manuf Technol 101:773–783. https://doi.org/10.1007/s00170-018-2934-5

Madanchi N, Thiede S, Gutowski TG, Herrmann C (2019b) Modeling the impact of cutting fluid strategies on environmentally conscious machining systems. In: 26th CIRP international conference on life cycle engineering, vol 80, pp 150–155, West Lafayette, USA. https://doi.org/10.1016/j.procir.2019.01.068

Malkin S, Guo C (2008) Grinding technology: theory and applications of machining with abrasives, 2nd edn. Industrial Press, New York, USA

Mang T, Dresel W (2007) Lubricants and lubrication. Wiley-VCH Verlag, Weinheim, Germany

Marinescu ID, Hitchiner MP, Uhlmann E, Rowe WB, Inasaki I (2007) Handbook of machining with grinding wheels. CRC Press Taylor & Francis Group, Boca Raton, USA

Mativenga PT, Rajemi MF (2011) Calculation of optimum cutting parameters based on minimum energy footprint. CIRP Ann 60:149–152. https://doi.org/10.1016/j.cirp.2011.03.088

Meister M (2011) Vademecum des Schleifens. Hanser, München, Germany

Metzger JL (1986) Superabrasive grinding. Butterworths, London, UK

Morgan MN, Jackson AR, Wu H, Baines-Jones V, Batako A, Rowe WB (2008) Optimisation of fluid application in grinding. CIRP Ann 57:363–366. https://doi.org/10.1016/j.cirp.2008.03.090

Nalbant M, Gökkaya H, Sur G (2007) Application of Taguchi method in the optimization of cutting parameters for surface roughness in turning. Mater Des 28:1379–1385. https://doi.org/10.1016/j.matdes.2006.01.008

Petuelli G (ed) (2002) Simulation des Kühlschmierstoffkreislaufs zur Optimierung einer umwelt- und ressourcenschonenden Produktionstechnik. DBU Forschungsbericht, FZ 13581. Shaker, Aachen, Germany

Pfleiderer C (1961) Die Kreiselpumpen für Flüssigkeiten und Gase. Springer, Berlin, Heidelberg, Germany

Priarone PC, Robiglio M, Settineri L, Tebaldo V (2016) Modelling of specific energy requirements in machining as a function of tool and lubricoolant usage. CIRP Ann 65:25–28. https://doi.org/10.1016/j.cirp.2016.04.108

Pusavec F, Kramar D, Krajnik P, Kopac J (2010) Transitioning to sustainable production—part II: evaluation of sustainable machining technologies. J Clean Prod 18:1211–1221. https://doi.org/10.1016/j.jclepro.2010.01.015

Rief M (2012) Vorhersagemodell für den Energiebedarf bei der spanenden Bearbeitung für eine energieeffiziente Prozessgestaltung. Dr.-Ing. Dissertation, Universität Magdeburg. Shaker, Aachen, Germany

Riegert W, Mager K (2001) Entwicklung einer Kleinzentrifuge zur dezentralen Trocknung von mit Kühlschmierstoffen behafteten Spänen. Abschlußbericht ARP, Alpirsbach-Peterzell, Germany

Rowe WB (2009) Principles of modern grinding technology. William Andrew, Oxford, UK

Sanchez LEdA, Palma GL, Marinescu I, Modolo DL, Nalon LJ, Santos AE (2013) Effect of different methods of cutting fluid application on turning of a difficult-to-machine steel (SAE EV-8). Proc Inst Mech Eng, Part b: J Eng Manuf 227:220–234. https://doi.org/10.1177/0954405412467589

Sangermann H (2013) Hochdruck-Kühlschmierstoffzufuhr in der Zerspanung. Dr.- Ing. Dissertation, RWTH Aachen. Apprimus Verlag, Aachen, Germany

Sangwan KS, Sihag N (2019) Multi-objective optimization for energy efficient machining with high productivity and quality for a turning process. In: 26th CIRP international conference on life cycle engineering, vol 80, pp 67–72, West Lafayette, USA. https://doi.org/10.1016/j.procir.2019.01.022

Schlosser R (2013) Methodik zur Prognose der Nachhaltigkeit des Energie- und Stoffeinsatzes spanender Fertigungsprozesse. Dr.-Ing. Dissertation, RWTH Aachen. Apprimus Verlag, Aachen, Germany

Schönherr H (2002) Spanende Fertigung. Oldenbourg, München, Germany

Schrems S (2014) Methode zur modellbasierten Integration des maschinenbezogenen Energiebedarfs in die Produktionsplanung. Dr.-Ing. Dissertation, TU Darmstadt. Shaker Verlag, Aachen, Germany

Sigloch H (2008) Technische Fluidmechanik, 6th edn. Springer Verlag, Berlin, Heidelberg, Germany

Songmene V, Zaghbani I, Kientzy G (2018) Machining and machinability of tool steels: effects of lubrication and machining conditions on tool wear and tool life data. In: 8th CIRP conference on high performance cutting, vol 77, pp 505–508. https://doi.org/10.1016/j.procir.2018.08.252

Ståhl J-E, Schultheiss F, Hägglund S (2011) Analytical and experimental determination of the ra surface roughness during turning. In: 1st CIRP conference on surface integrity, procedia engineering, vol 19, pp 349–356. https://doi.org/10.1016/j.proeng.2011.11.124

Su H, Yang C, Gao S, Fu Y, Ding W (2019) A predictive model on surface roughness during internal traverse grinding of small holes. Int J Adv Manuf Technol 103:2069–2077. https://doi.org/10.1007/s00170-019-03643-z

Tornau D, Bradke H-J, Rahe W, Schühly H-P (1999) Kühlschmierstoffe und Anlagen: Theorie und Praxis. Kontakt & Studium, vol 579. expert verlag, Renningen-Malmsheim, Germany

Torrance AA, Badger JA (2000) The relation between the traverse dressing of vitrified grinding wheels and their performance. Int J Mach Tools Manuf 40:1787–1811. https://doi.org/10.1016/S0890-6955(00)00015-8

Triebs JB (2015) Analyse, Optimierung und Simulation des Energieumsatzes dezentraler Fluidkreisläufe von Werkzeugmaschinen. Dr.-Ing. Dissertation, RWTH Aachen. Apprimus Verlag, Aachen, Germany

Triesch J (2010) Reinigung von Kühlschmierstoffen: Konzepte, Methoden und Hinweise für den Praktiker, 1st edn. expert verlag, Renningen, Germany

VDI 2055 (2019) Thermal insulation of heated and refrigerated operational installations. Verein Deutscher Ingenieure, Berlin, Germany

VDI 2089 (2010) Building services in swimming baths indoor pools. Verein Deutscher Ingenieure, Berlin, Germany

VDI 3035 (2008) Design of machine tools, production lines and peripheral equipment for the use of metalworking fluids. Verein Deutscher Ingenieure, Berlin, Germany

VDI 3321 (1994) Schnittwertoptimierung Grundlagen und Anwendung. Verein Deutscher Ingenieure, Berlin, Germany

VDI 3802-2 (2012) Air conditioning systems for factories. Verein Deutscher Ingenieure, Berlin, Germany

Victor H (1969) Schnittkraftberechnungen für das Abspanen von Metallen. Wt. Zeitschrift Für Industrielle Fertigung 59:311–326

Wang Q, Liu F, Wang X (2014) Multi-objective optimization of machining parameters considering energy consumption. Int J Adv Manuf Technol 71:1133–1142. https://doi.org/10.1007/s00170-013-5547-z

Wang R, Wang B, Barber G, Gu J, Schall J (2019) Models for prediction of surface roughness in a face milling process using triangular inserts. Lubricants 7:9. https://doi.org/10.3390/lubricants7010009

Wilbert H-P, Kreisel G, Goldhan G (1998) Ganzheitliche Bilanzierung/Bewertung von Reinigungs- / Vorbehandlungstechnologien in der Oberflächenbehandlung: Forschungsbericht Ganzheitliche Bilanzierung / Fallstudien. Bundesministerium für Bildung und Forschung (BMBF)

Winter M (2016) Eco-efficiency of Grinding Processes and Systems. Dr.-Ing. Dissertation, Technische Universität Braunschweig. Springer, Switzerland

Winter M, Thiede S, Herrmann C (2015) Influence of the cutting fluid on process energy demand and surface roughness in grinding - a technological, environmental and economic examination. Int J Adv Manuf Technol 77:2005–2017. https://doi.org/10.1007/s00170-014-6557-1

Xie Y, Williams JA (1996) The prediction of friction and wear when a soft surface slides against a harder rough surface. Wear 196:21–34

Yan J, Li L (2013) Multi-objective optimization of milling parameters—the trade-offs between energy, production rate and cutting quality. J Clean Prod 52:462–471. https://doi.org/10.1016/j.jclepro.2013.02.030

Yegenoglu K (1986) Berechnung von Topographiekenngrößen zur Auslegung von CBN-Schleifprozessen. Dr.-Ing. Dissertation, RWTH Aachen, Aachen, Germany

Zein A (2012) Transition towards energy efficient machine tools. Dr.-Ing. Dissertation, Technische Universität Braunschweig. Springer, Berlin, Germany

Zein A, Li W, Herrmann C, Kara S (2011) Energy efficiency measures for the design and operation of machine tools: an axiomatic approach. In: 18th CIRP international conference on life cycle engineering, Braunschweig, Germany 274–279. https://doi.org/10.1007/978-3-642-19692-8_48

Zitt UR (1999) Modellierung und Simulation von Hochleistungsschleifprozessen. Dr.-Ing. Dissertation, Universität Kaiserslautern, Kaiserslautern, Germany

Züst S, Gontarz A, Wegener K (2013) Energy equivalent of compressed air consumption in a machine tool environment. In: 11th global conference on sustainable manufacturing, pp 359–364

Züst SD (2017) Model based optimization of internal heat sources in machine tools. Dr. sc. Dissertation, ETH Zurich. Zurich, Switzerland

Chapter 5
Exemplary Application of the Concept

In this chapter the developed concept is applied to different application cases in order to underline its applicability and potential. By using the implemented concept according to the previous chapter, individual machining systems are analyzed and improved (Sect. 5.1). Moreover, the planning of multiple machining systems within a process chain is also considered and the impact of specific improvement measures is demonstrated (Sect. 5.2). In addition, an overview of all considered measures in relation to their area of application is presented (Sect. 5.3).

5.1 Application to a Machining System

For the first part of the concept application three individual machining systems are independently analyzed. As presented in Fig. 5.1, these comprise a turning, a milling and a grinding process including the peripheral subsystems as well as a subsequent cleaning process and the scrap treatment. The developed concept is applied for each machining system individually by using the implemented software tool following the framework presented in Fig. 4.32.

5.1.1 Analyzing and Evaluating a Machining System

Initially, the scope of the analysis is determined. The case study is conducted from a production planning point of view, where the stakeholder aims at evaluating the environmental, economic and technological performance of an existing machining system and at improving it. For this analysis, each machining system is used for one specific production step to produce an exemplary workpiece made of steel with

N. Madanchi, *Model Based Approach for Energy and Resource Efficient Machining Systems*, Sustainable Production, Life Cycle Engineering and Management,
https://doi.org/10.1007/978-3-030-87540-4_5

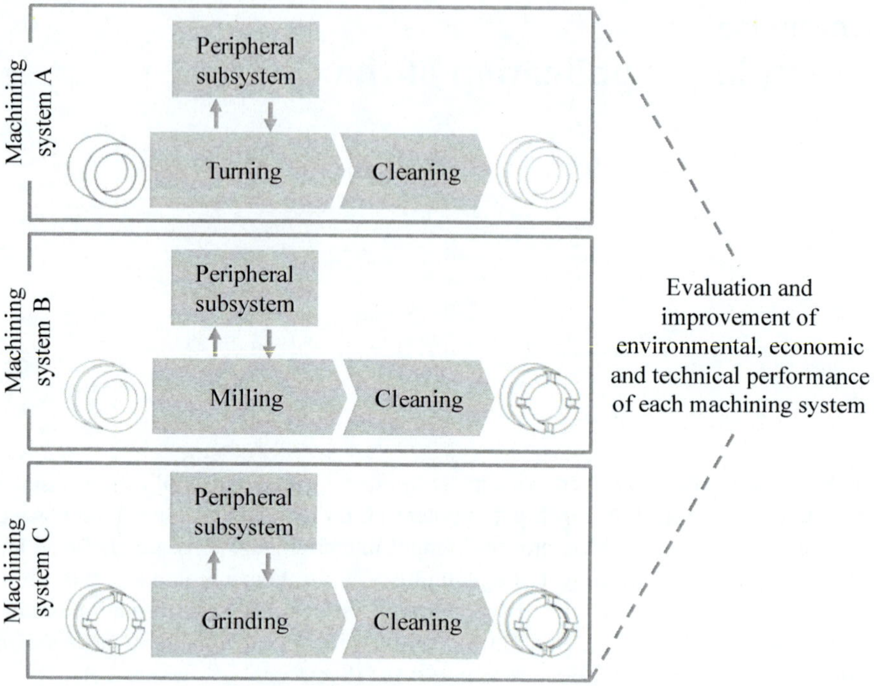

Fig. 5.1 Considered machining systems for the concept application

the designation 1.1191 (C45) as demonstrated in Fig. 5.1. In case of turning, the functional unit describes the production of one workpiece by removing 50265 mm^3 of material during external turning. Similarly, the milling process considers the material removal of 5760 mm^3 and the internal grinding process a workpiece volume of 100 mm^3. Further, the production profile is described. A general standby rate of 20% is assumed due to e.g. waiting or maintenance. Detailed values for the used labor and machine tool usage cost rates as well as others are listed in the appendix. Following the concept, the process and workpiece parameters as well as the used tools are specified in more detail. The turning process is conducted by using cemented carbide inserts with four possible cutting edges. On the basis of the used tool and process settings, the recommended average process parameters are used with a cutting speed of 150 m/min, cutting depth of 1 mm and a feed rate of 0.2 mm/rev. Similarly, the milling process is conducted by using a cemented carbide end mill with a diameter of 10 mm and a cutting speed of 90 m/min, cutting depth of 1 mm and a feed rate per cutting edge of 0.04 mm/rev. The internal grinding process is carried out by using a CBN grinding wheel with a diameter of 40 mm and a width of 15 mm. A cutting speed of 60 m/s, cutting depth of 0.0015 mm and a workpiece speed of 0.5 m/s are used. For grinding, it is further required to specify the dressing process with a dressing depth of 0.001 mm and a traverse dressing speed of 374.54 mm/min.

For this case study, the flooding strategy by using a mineral oil-based emulsion with a concentration of 7.5% is analyzed. Regarding the peripheral subsystems, it is further specified that each machining system is supplied by a decentralized cutting fluid and exhaust air system. It is assumed that the cutting fluid system is completely enclosed and has a capacity of 500l. The system uses two pumps, where one pump supplies the fluid to the machine tool into the contact zone and the other pump is used to circulate and filter the pump within a by-pass (see Fig. 2.11). Therefore, the supply pump is only active during processing and dressing, while the circulation pump is always active during production time. Similarly, the exhaust air system is also active during production time. The cutting fluid return and chip transportation is designed as a gravity return flow without the need for an additional pump back station. In order to remove cutting fluid from the scrap material, a de-oiling process is considered for each machining system using a centrifuge with a capacity of 2 kg/min. It is assumed that the removed cutting fluid is not reused and lost to the system. Following the machining process, the workpiece is treated by an immersion cleaning process. For this process, a cleaning fluid temperature of 70 °C is assumed. The cleaning system is further considered to be enclosed and is only opened to remove the workpieces. The machining systems run mainly automatically and one worker can thus operate three machine tools simultaneously.

After the specification of the machining system and data input, the models that are implemented into the software tool are used to determine the energy and resource flows and to calculate the environmental, economic and technological impact. The values for the CO_2-equivalent and costs are taken from literature and are also presented in the appendix. These results of the current situation are used as the base scenario and possible changes or improvements for the production planning are evaluated in comparison to it. The results for the base scenario regarding the environmental impact are demonstrated in Fig. 5.2 for each machining system.

It demonstrates the highest environmental impact for the milling system. Compared to grinding and turning it is striking that the impact due to the tool is significantly higher. Although the tool wear is even slightly less for milling than for turning, the impact is caused by the significantly higher CO_2-equivilent per tool. While the insert used as turning tool weighs 9.5 g with four usable cutting edges, the end mill for milling weighs 87 g, which leads to the different CO_2-equivilents. However, for grinding the share of grinding wheel wear to the overall usable wheel is very low, resulting in a low environmental impact despite the comparatively high CO_2-equivilent for the grinding wheel itself.

For all three machining systems, the environmental impact due to the direct use of energy is particularly relevant. Regarding the milling process the impact within the actual processing state is comparatively lower, which is mainly caused by the shorter cutting time. Higher cutting times also lead to more heat input into the cutting fluid due to the supply pump and the process and is thus noticeable by the higher cooling unit impact for grinding and turning.

In comparison, the machine tool used for milling has the largest base load. This is a main aspect for the higher environmental impact during standby and preparation

Turning:		**Milling:**		**Grinding:**	
Cutting speed:	v_c = 150 m/min	Cutting speed:	v_c = 120 m/min	Cutting speed:	v_c = 60 m/s
Cutting depth:	a_e = 1.0 mm	Cutting depth:	a_e = 1.0 mm	Workpiece speed:	v_w = 0.5 m/s
Feed rate:	f = 0.2 mm/rev	Feed rate:	f_z = 0.04 mm/rev	Cutting depth :	a_e = 1.5 µm
Removed volume:	V_w = 50265 mm³	Removed volume:	V_w = 5760 mm³	Removed volume:	V_w = 100 mm³

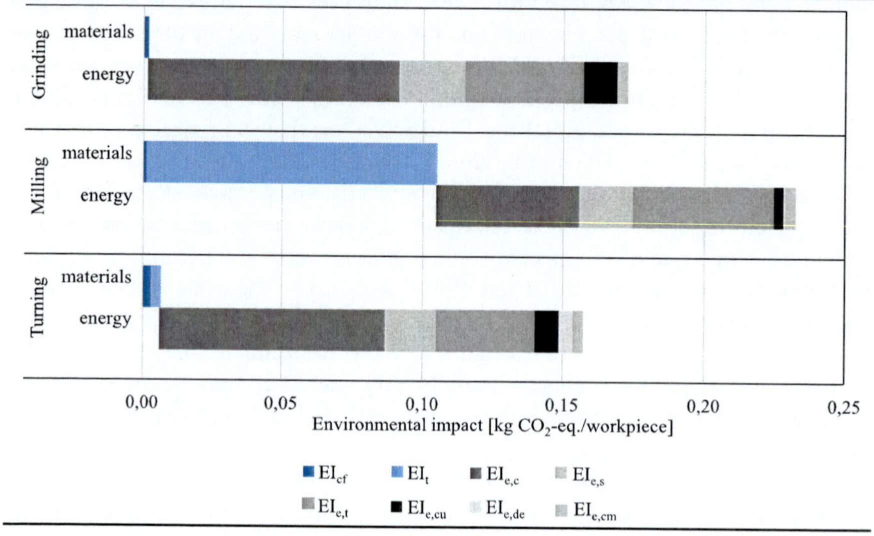

Fig. 5.2 Environmental performance of machining systems

state. Further, the preparation time for workpiece and tool is rather high compared to the cutting or standby time.

The workpiece material removal is significantly higher for turning than for the other two machining processes. This amount of scrap material leads to a larger impact of the de-oiling process and also more cutting fluid drag-out and thus a slightly larger cutting fluid impact.

Similarly, the economic impact of the three machining systems are presented in Fig. 5.3. Here, the impact is demonstrated for the demand of energy, tool, cutting fluid and the system. The system comprises the use of the machine tool, cleaning and de-oiling system as well as required labor.

Overall, the milling system demonstrates the highest economic impact. This is mainly related to the significant influence of the end mill. Although the price for a grinding wheel is clearly more expensive, the economic impact is almost negligible for the grinding system due to the large tool life. This is the opposite for the energy demand. Despite the intensive consumption, the price for electrical energy is rather low and thus also the impact. Regarding the cutting fluid usage the consumption as well as the price for mineral oil based emulsion is also rather low.

Overall, the analysis of the economic impact clearly demonstrates the strongest influence due to system usage. The high costs for labor and machine usage over time lead to this significant impact for each machining system.

Turning:		**Milling:**		**Grinding:**	
Cutting speed:	$v_c = 150$ m/min	Cutting speed:	$v_c = 120$ m/min	Cutting speed:	$v_c = 60$ m/s
Cutting depth:	$a_e = 1.0$ mm	Cutting depth:	$a_e = 1.0$ mm	Workpiece speed:	$v_w = 0.5$ m/s
Feed rate:	$f = 0.2$ mm/rev	Feed rate:	$f_z = 0.04$ mm/rev	Cutting depth :	$a_e = 1.5$ µm
Removed volume:	$V_w = 50265$ mm³	Removed volume:	$V_w = 5760$ mm³	Removed volume:	$V_w = 100$ mm³

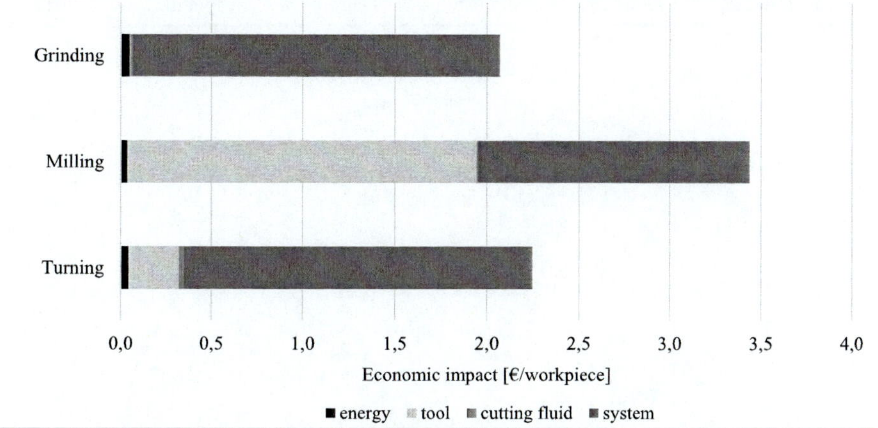

Fig. 5.3 Economic performance of machining systems

5.1.2 Impact of Process Parameters and Cutting Fluid Strategies

In order to analyze the impact of changing process parameters or cutting fluid strategies, a simplified sensitivity analysis has been conducted for this use case. For this approach the presented base scenario is used as a reference regarding the settings and results. In accordance with the general assumption that cutting speed, feed rate and cutting depth are particularly relevant (VDI 3321 1994), these process parameters are varied as well as the cutting fluid strategy. While one parameter is changed the others remained at the reference setting. In order to enable a consistent approach and a comparison, each reference value is changed by ±50% to obtain a minimal and maximal value. The impact of changed parameters and strategies are analyzed for the production time, consumption of materials by cutting fluid and tool usage, direct energy consumption of each subsystem as well as the environmental, economic and technological performance indicators. For this analysis, the results of the base scenario are considered as 1 and the changed results are normalized against it. The results of the analysis are presented in Fig. 5.4 for the turning system, in Fig. 5.5 for the milling system and in Fig. 5.6 for the grinding system.

For the turning system, this analysis demonstrates that the production time is equally influenced by the process parameters, which determine the material removal rate and thus the cutting time. However, regarding the consumption of cutting fluid, the choice of strategy and especially dry turning has the strongest impact. The consumption of the cutting tool by contrast is dominated by the impact of the cutting

Reference setting:		**Changed setting :**	
Cutting speed:	v_c = 150 m/min	Cutting speed:	v_c = 75; 225 m/min
Cutting depth:	a_e = 1.0 mm	Cutting depth:	a_e = 0.5; 1.5 mm
Feed rate:	f = 0.2 mm/rev	Feed rate:	f = 0.1; 0.3 mm/rev
CF strategy:	Flooding, 7.5 % emulsion	CF strategy:	Flooding, oil and dry

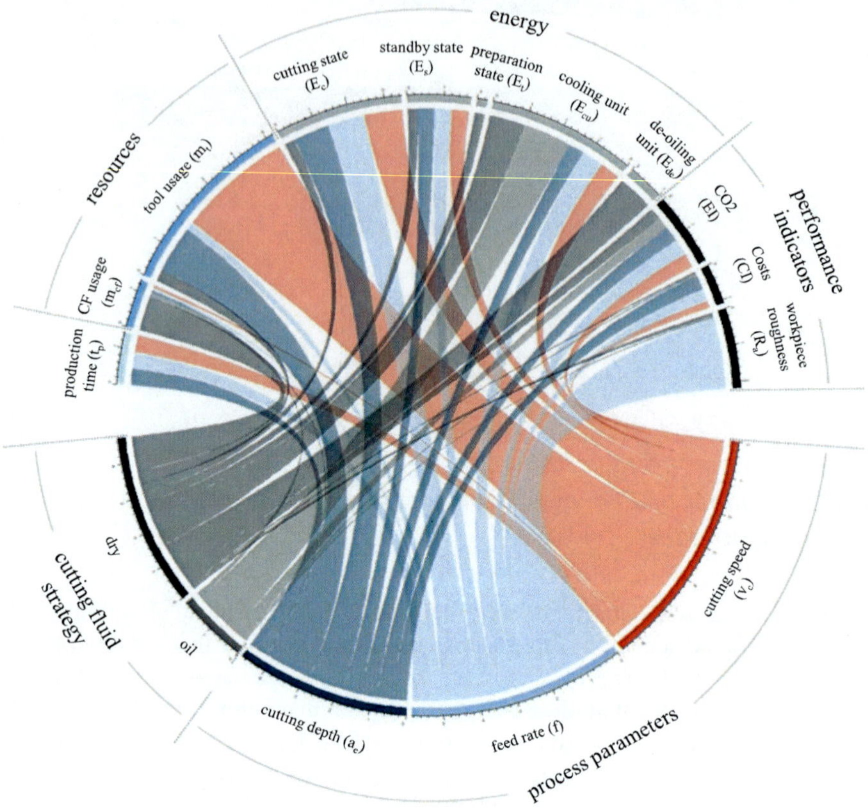

Fig. 5.4 Impact of changing parameters and strategies on the turning system

speed. Regarding the energy consumption of the subsystems it is striking that the preparation state as well as the de-oiling unit are only influenced by the strategy of dry turning. In that case, the entire de-oiling process is eliminated and the base load of the system is reduced by the cutting fluid system during the preparation state. This is also true for the standby and cutting state, but here the process parameters are also relevant. This also applies for the cooling unit, but in this case the strongest impact was demonstrated by the strategy to use oil instead of an emulsion. Due to the changed fluid properties, the cooling and thus energy demand increased significantly. With regard to the performance indicators the analysis reveals that the environmental performance is nearly equally influenced by process parameters and cutting fluid strategy, while the economic performance demonstrates a stronger impact of the process parameters. As shown the process parameters determine the production time and therefore also the labor and machine usage, which are particularly important in terms of costs. The impact on the technological performance, represented as surface roughness is dominated by the feed rate.

Overall, the results for the milling system are rather similar to the turning system. However, as a difference it can be noted that generally the feed rate demonstrates the strongest impact in case of milling, whereas this applies to cutting speed in case of turning. Further, the consumption of the tool is not dominated by the cutting speed and

Reference setting:		**Changed setting :**	
Cutting speed:	$v_c = 120$ m/min	Cutting speed:	$v_c = 60$; 180 m/min
Cutting depth:	$a_e = 1.0$ mm	Cutting depth:	$a_e = 0.5$; 1.5 mm
Feed rate:	f = 0.04 mm/rev	Feed rate:	f = 0.02; 0.06 mm/rev
CF strategy:	Flooding, 7.5 % emulsion	CF strategy:	Flooding, oil and dry

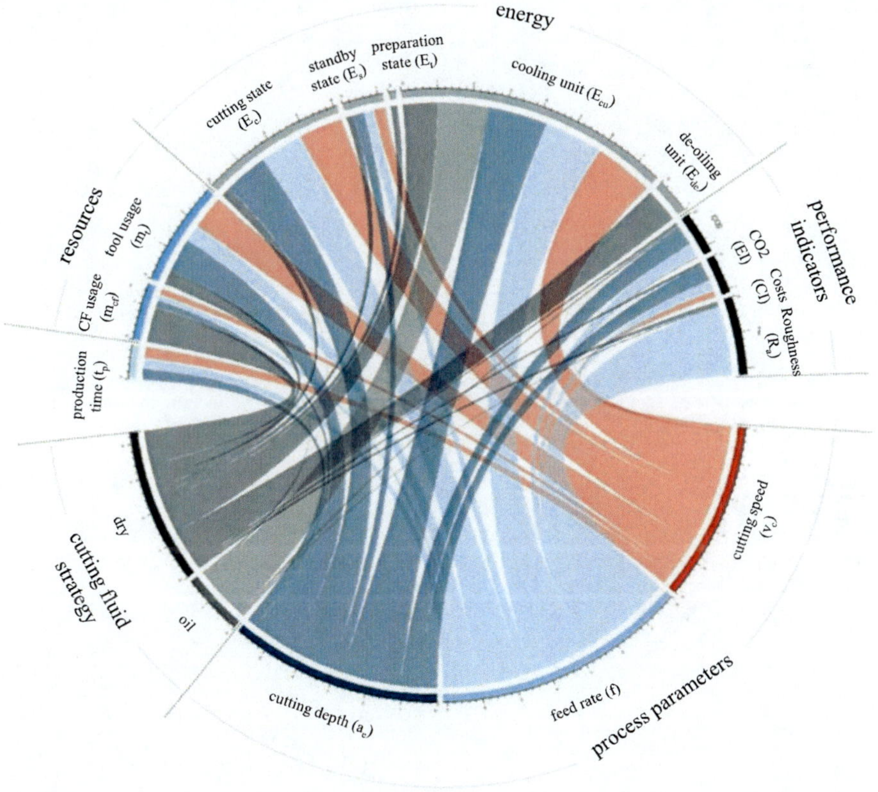

Fig. 5.5 Impact of changing parameters and strategies on the milling system

rather depends on the cutting depth. Also, the performance indicators demonstrate a greater influence due to the cutting depth and feed rate compared to the cutting speed.

The results of the analysis differ more strongly for the grinding system. Overall, they reveal that the workpiece speed has the strongest impact. The cutting speed by contrast, only has a strong impact on the consumption of the grinding wheel and the surface roughness of the workpiece. As the cutting speed is not used to determine the material removal rate for grinding, other results such as production time and energy consumption are only slightly influenced by it. In comparison to the other machining systems, it can further be noted that the energy consumption during the

Reference setting:		**Changed setting :**	
Cutting speed:	$v_c = 60$ m/s	Cutting speed:	$v_c = 30; 90$ m/s
Cutting depth:	$a_e = 1.0$ µm	Cutting depth:	$a_e = 0.5; 1.5$ µm
Workpiece speed:	$v_w = 0.5$ m/s	Workpiece speed:	$v_w = 0.25; 0.75$ m/s
CF strategy:	Flooding, 7.5 % emulsion	CF strategy:	Flooding, oil

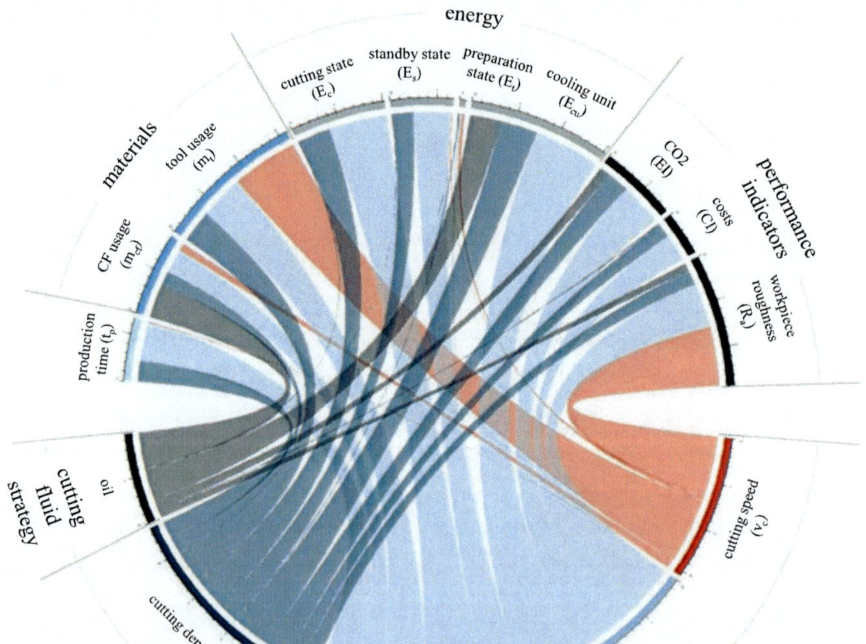

Fig. 5.6 Impact of changing parameters and strategies on the grinding system

preparation state is also influenced by the process parameters. This is caused by the consideration of the dressing process. Within this analysis, only the strategy of using a grinding oil was investigated for the grinding system. The use of this strategy mainly influences the results regarding cutting fluid consumption, cooling unit, environmental performance and surface roughness.

5.1.3 Application of Improvement Measures

The developed concept also supports the improvement of the overall performance of the machining systems. For this purpose, different organizational and technical improvement measure were presented in Sect. 4.4. The focus here is on the selection of alternative process parameters and cutting fluid strategies.

As the concept is designed to analyze and to evaluate the impact of cutting fluid strategies and process parameters on machining systems, these measure are applied to the presented turning, milling and grinding system. For each cutting fluid strategy the selection of the appropriate process parameters differs, hence both measures are considered in combination for this case study. Compared to the analysis from the previous section, which already demonstrated the relative impact of different strategies and parameters, this analysis focuses on the absolute results and actual values of the performance indicators.

For the turning process, the strategies of using mineral oil based emulsion, oil and dry turning are considered. For each strategy, the following exemplary objective function and constraints are used based on the limitations imposed by the used tool and material:

$$Z = \mathrm{Min}(EI_{MS},\ CI_{MS})$$

Subject to

$$\begin{aligned} 75 \leq v_c \leq 225 \\ 0.1 \leq f \leq 0.3 \\ 0.5 \leq a_e \leq 1.5 \\ R_a \leq 3.0 \end{aligned} \tag{5.1}$$

The model is solved by using the solver in MS Excel®. For the application, it is assumed that the overall environmental and economic indicators are to be minimized, while the technological indicator represented by the arithmetic mean roughness of the workpiece is considered as a constraint. As turning is not considered as a finishing process, it is usually not the aim to also minimize the surface roughness, but it can be included in the objective function if required. The environmental and economic indicators are equally weighted and normalized against the base scenario. This results into different process parameters with varying cutting times for each cutting fluid strategy as presented in Fig. 5.7.

Similarly, the milling process also focuses on the same three cutting fluid strategies in combination with the process parameters. For this use case, the objective function and exemplary constraints are formulated as follows:

$$Z = \mathrm{Min}(EI_{MS},\ CI_{MS})$$

Subject to

$$\begin{aligned} 60 &\le v_c \le 180 \\ 0.02 &\le f_z \le 0.06 \\ 0.5 &\le a_e \le 1.5 \\ R_a &\le 0.5 \end{aligned} \tag{5.2}$$

It is again assumed that only the overall environmental and economic indicators are to be minimized with equal weighting of importance. The arithmetic mean roughness of the workpiece, however, is considered as a constraint. The results of the combination between cutting fluid strategy and process parameters are presented in Fig. 5.8.

Emulsion		**Oil**		**Dry**	
Cutting speed:	v_c = 173 m/min	Cutting speed:	v_c = 174 m/min	Cutting speed:	v_c = 158 m/min
Cutting depth:	a_e = 1.5 mm	Cutting depth:	a_e = 1.5 mm	Cutting depth:	a_e = 1.5 mm
Feed rate:	f = 0.2 mm/rev	Feed rate:	f = 0.21 mm/rev	Feed rate:	f = 0.19 mm/rev
Cutting time:	t_c = 57 s	Cutting time:	t_c = 55 s	Cutting time:	t_c = 75 s

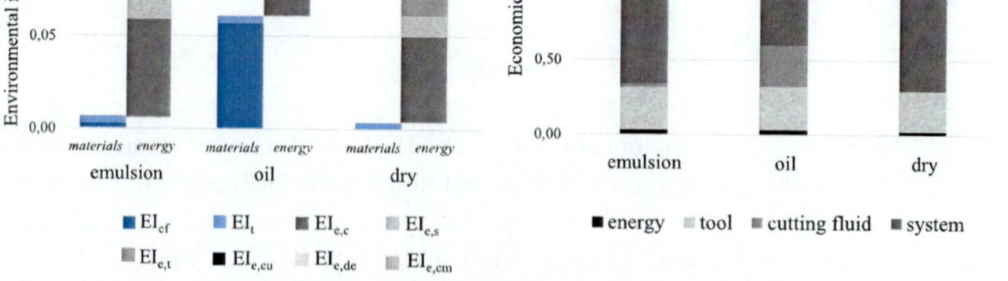

Fig. 5.7 Results of alternative strategies and parameters for the turning system

Emulsion		Oil		Dry	
Cutting speed:	$v_c = 79.7$ m/min	Cutting speed:	$v_c = 79.4$ m/min	Cutting speed:	$v_c = 61.1$ m/min
Cutting depth:	$a_e = 1.5$ mm	Cutting depth:	$a_e = 1.5$ mm	Cutting depth:	$a_e = 1.5$ mm
Feed rate:	$f_z = 0.04$ mm/rev	Feed rate:	$f_z = 0.04$ mm/rev	Feed rate:	$f_z = 0.04$ mm/rev
Cutting time:	$t_c = 55$ s	Cutting time:	$t_c = 53$ s	Cutting time:	$t_c = 75$ s

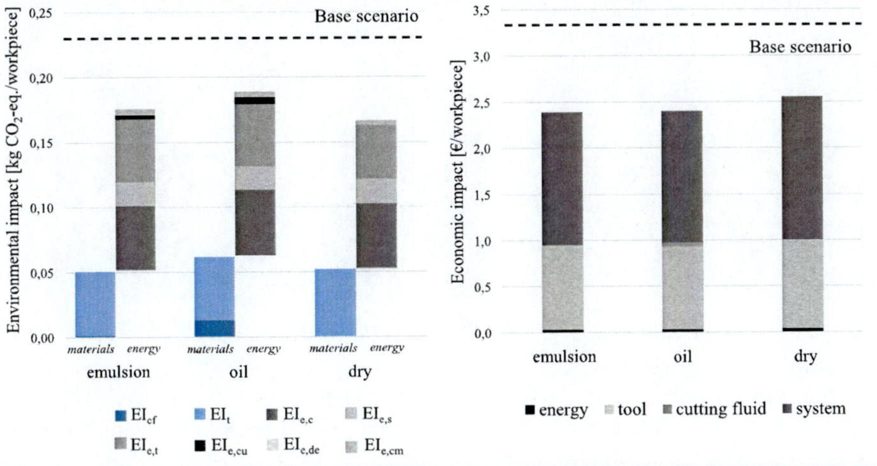

Fig. 5.8 Results of alternative strategies and parameters for the milling system

With regard to the grinding system only the two cutting fluid strategies of using mineral oil based emulsion and mineral oil are taken into account. As grinding is considered as a finishing process with the aim to create workpieces with high surface quality, the exemplary objective function for grinding includes the arithmetic mean roughness of the workpiece.

$$Z = \mathrm{Min}(EI_{MS},\ CI_{MS},\ R_a)$$

Subject to

$$\begin{aligned} 30 &\leq v_c \leq 90 \\ 0.5 &\leq a_e \leq 1.5 \\ 0.25 &\leq v_w \leq 0.75 \end{aligned} \tag{5.3}$$

The importance of the three indicators is equally weighted for this case study and normalized against the base scenario. Figure 5.9 therefore presents the environmental and economic impact as well as the roughness in combination of the specific process parameters for both cutting fluid strategies.

Overall, the results demonstrate a significant improvement by the measures compared to the base scenario. An exception to this is the environmental impact due to oil usage for the turning and grinding system. For the turning system, the strategy of dry turning can generally be recommended within the context of this

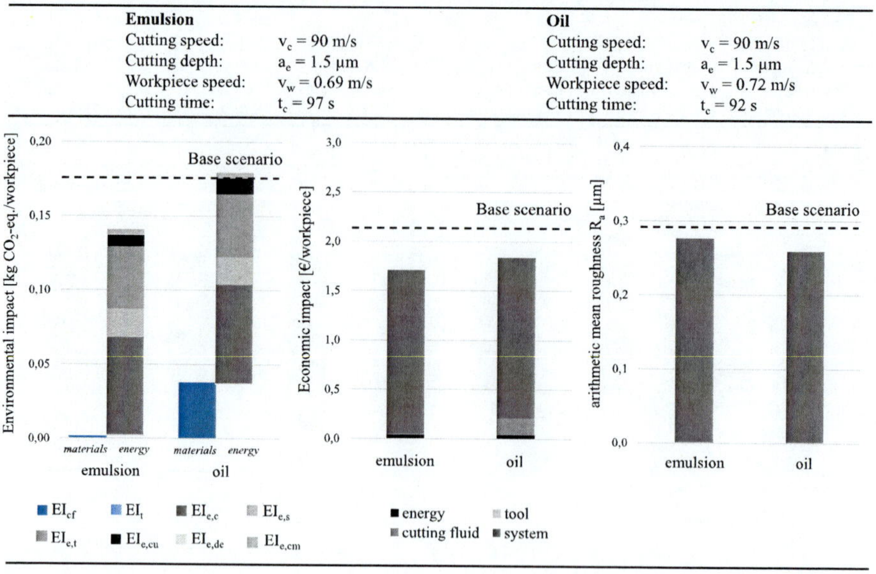

Fig. 5.9 Results of alternative strategies and parameters for the grinding system

case study. Although this strategy results in higher cutting times, the environmental impact is lower compared to the use of emulsion or oil. With regard to the economic impact, it is decisive that the costs for the cutting fluid system are eliminated, whereas the elimination of the de-oiling and cooling unit are also critical regarding the environmental impact. In addition, the selected process parameters for dry turning are not as demanding compared to the other two strategies and lead to less tool wear and cutting power but also to a lower material removal rate. Thus, if the material removal rate is also included as a constraint of the objective function, the results for this strategy are significantly reduced. The high impact for the use of oil mainly results from the impact of the fluid itself and the cooling unit.

The milling system, however, demonstrates different results. In this case, it could be recommended to use the emulsion. Due to the high costs and CO_2-equivilents for the end mill, the tool wear and usage is particularly relevant for milling. This causes lower process parameters for dry milling with increased cutting time and thus system usage. As a consequence the economic impact is also increased, whereas the reduced tool wear and cutting power again reduce the environmental impact as well.

For the grinding system, only the flooding strategies are considered and in contrast to the other two machining systems the improvement measure simultaneously aims at minimizing the environmental, economic and technological impact. The results clearly demonstrate a higher environmental impact by using oil due to the impact of the fluid and higher energy demand of the cooling unit. For the economic impact, this difference is not as obvious. In this case, the oil usage causes lower cutting times and thus use of the system. The selected process parameters also lead to slightly lower values for the arithmetic mean surface roughness.

In general, the results demonstrate that the combination of cutting fluid strategy and process parameters can lead to significant improvements for the machining system and that it has to be selected individually depending on the specific constraints of the system and use case.

5.1.4 Demonstration of Further Possible Measures

In addition to the selection of alternative process parameters and cutting fluid strategies, further organizational and technical improvement measures can be applied to improve the performance of individual machining systems. Although these measures are currently not modeled within the concept, their potential and impact is experimentally demonstrated.

Emission reduction

In order to reduce the emissions resulting from machining processes, a technical measure concerning the exhaust air system is investigated within the grinding process of the case study. This measure aims at the shielding of cutting fluid emissions generated during machining. It is supposed to shield the emissions around the machining process directly inside the machine tool and thereby reducing the drag-out of cutting fluid emissions and the necessary exhaust air volume flow. The importance of cutting fluid emissions and relevance regarding occupational health and safety has already been presented in Sect. 2.3.3. The measure is implemented by generating a bell-shaped film of cutting fluid, which shields the machining process from the surrounding area and thus reduces the work space. The generation of the bell-shaped fluid film is based on the principle of bell fountains. Especially, small particles and vapors should not pass through the closed cutting fluid film. Emissions can further condense on the inside of the fluid film and automatically become part of the cutting fluid system again. Although the large surface of the fluid bell may also emit vapors, this is rather negligible compared to the amount of cutting fluid emissions generated by the machining process.

In order to demonstrate the effectiveness of this measure it is applied to the internal grinding process of the presented use case. As presented in Fig. 5.10 the grinding process is performed within the bell-shaped cutting fluid film, while the particles that are still generated, are measured using the Handheld 3016® from Lighthouse. This device measures particles with a size from 0.3 to 25 μm and classifies them into six groups.

The results of the experimental investigation are presented in Fig. 5.11. The demonstrated distribution of particles generally confirms the statement that an increasing cutting speed also increases the number of particles. Although the used measuring device has a lower resolution, it can also be verified that the highest share of particles is at a smaller size. With regard to the implemented improvement measure, the results demonstrate a positive impact. In relation, especially larger particles and particles at higher cutting speeds are reduced. Overall, an average particle reduction

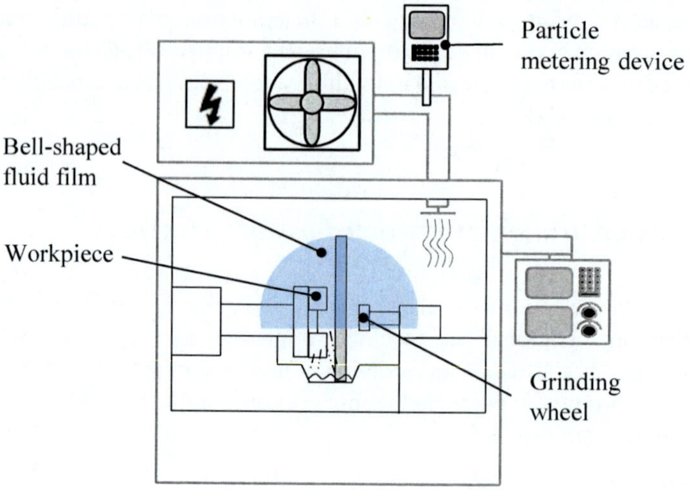

Fig. 5.10 Experimental setup for the implementation of emission reduction

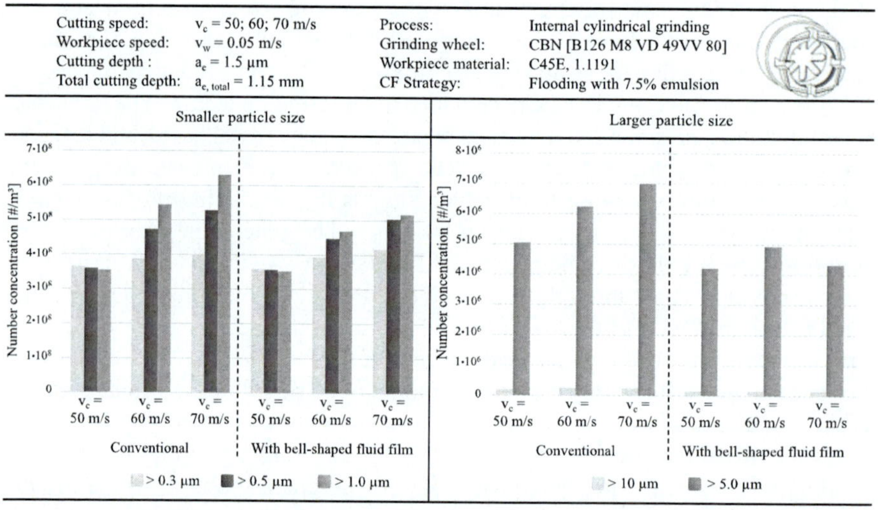

Fig. 5.11 Impact of a bell-shaped fluid film on generation of particles

of 15% can be observed. However, as stated by Madanchi et al. (2019) particles with a size of 2 µm and higher are particularly relevant in terms of occupational health and drag-out of cutting fluid. For these particle sizes, the reduction is even around 30%. Resulting from the reduction of cutting fluid particles it could also be possible to reduce the exhaust air volume flow and to lower its energy consumption. However, for the application of this measure it is important that there is no turbulence of the fluid film by e.g. contact with the rotating tool or workpiece. This leads to the generation of additional emissions and thus creates the opposite effect.

Nozzle design

Another technical measure focuses on the design of cutting fluid supply nozzles and its influence is experimentally presented. As demonstrated in Sect. 2.3.2.1 the nozzle design is particularly critical for grinding to deliver the cutting fluid into the actual contact zone. By influencing the cutting fluid speed and flow, the nozzle design and also nozzle position have to be adapted to overcome the air barrier circulating the grinding wheel. The measure aims at decreasing the required cutting fluid volume flow to transport a specific amount of useful cutting fluid volume flow into the contact zone. By using the flow separator and collector from Madanchi et al. (2017) presented in Fig. 5.12, this measure was exemplarily applied to a surface grinding process. Similar to other devices used in literature, this flow collector is used to capture only the cutting fluid passing through the contact zone, while the remaining volume flow is blocked out.

To demonstrate the application, experiments are carried out using a Blohm profimat 307 surface grinding machine with a maximal spindle power of 27 kW and spindle speed of 10,000 rev/min. Moreover, the cutting fluid supply system demonstrated in Fig. 4.21 with additional sensors to monitor the pressure, flow rate and temperature of the cutting fluid is used. The electrical power consumption of the cutting fluid system is monitored using a Chauvin Arnoux, C.A. 8335® metering device. For the application different flooding nozzles with a free jet are investigated and individually produced by using the 3D-printing technique fused deposition modeling. These nozzles are slot and round nozzles with varying outlet size and parallel length at the end of the nozzle. In order to compare the different nozzles, it is assumed that an average useful cutting fluid volume flow of 10 L/min ±0.5 is required for the grinding process. To realize this useful volume flow, different nozzle positions and supplied volume flows are individually required for each nozzle. The results of the experiments are demonstrated in Fig. 5.13.

At a comparative cutting fluid volume flow passing though the contact zone, the experimental results demonstrate that the required pump power varies for each nozzle

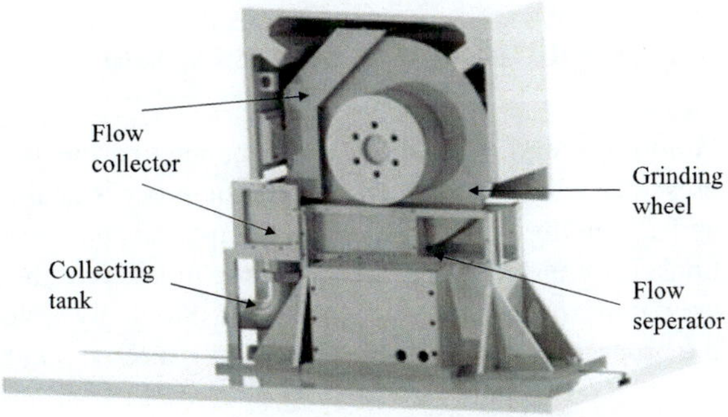

Fig. 5.12 Illustration of the flow separator and collector (Madanchi et al. 2017)

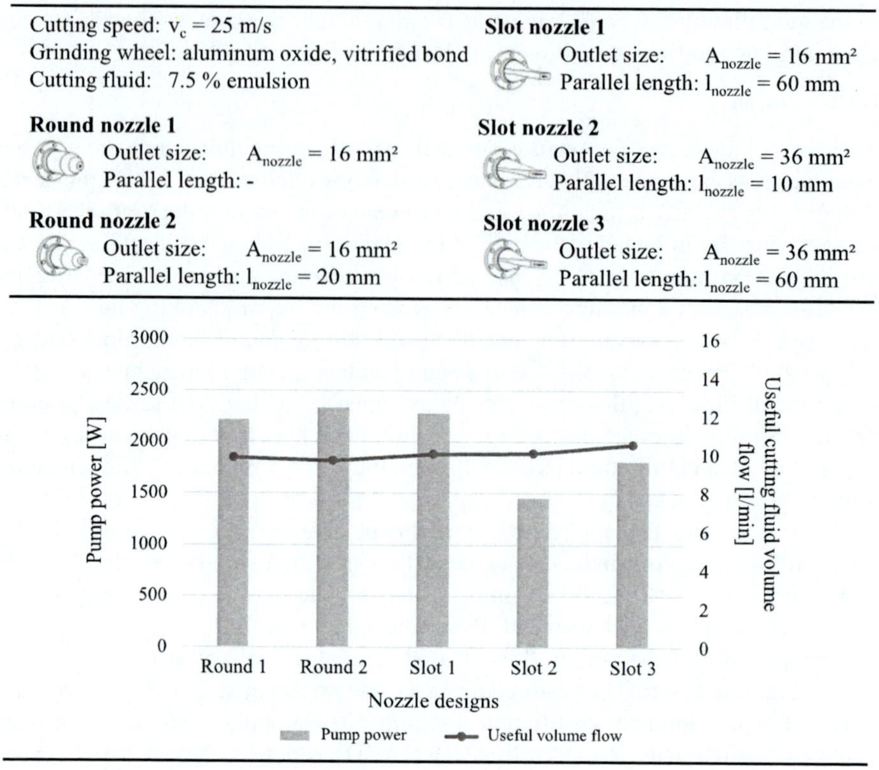

Fig. 5.13 Impact of cutting fluid nozzles on pump power

design. By using slot nozzle 2 instead of the round nozzle 2, the pump power can be reduced by around 900 W and thus 38%. As to be expected, the results from a technical perspective also demonstrated that the selected settings for each nozzle led to the same grinding forces and surface roughness of the workpiece within the case study.

5.2 Application to a Process Chain

For the second part of the concept application the machining process chain as presented in Fig. 5.14 is analyzed. The process chains considers a turning, a milling and a grinding process including peripheral subsystems as well as intermediate and final cleaning processes. Similar to the first part of the concept application, the developed concept and software tool is used to analyze and evaluate the performance of the process chain and improvement measures.

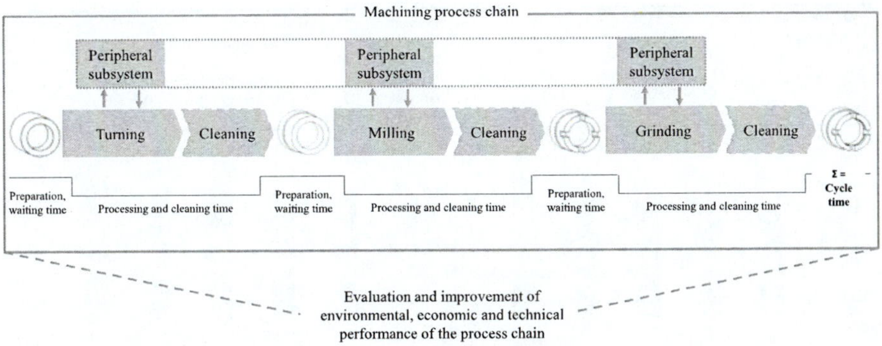

Fig. 5.14 Considered process chain for the concept application

5.2.1 Planning and Evaluating a Process Chain

Following the developed concept, the analysis starts by the definition of the scope. The purpose of analysis for this case study is the greenfield planning of a new machining process chain by comparing alternative configurations. For the process chain, three machining systems are required to produce a workpiece made of steel with the designation 1.1191 (C45) as considered in the previous section. To produce such a workpiece within a process chain the cycle time including processing, cleaning, preparation and waiting time are also relevant for the use case. The functional unit therefore describes the production of a workpiece with a cycle time of 20 min by external turning, milling and internal grinding. The production profile, initial process parameters and used cutting tools are also considered as previously described. For the planning of the process chain, the cutting fluid strategies of flooding using a mineral oil based emulsion, oil and dry machining are alternatively considered. With this regard, the alternatives of using centralized or decentralized peripheral subsystems are also evaluated. The size and capacity of these systems thus depend on the configuration. Further, the need for cleaning processes is considered based on the design of the process chain and the alternative use of surface air cleaners in the case of dry machining is also analyzed. In addition to these alternative strategies and configurations, further improvement measures targeting the standby operation and exchange of specific components with more efficient ones are also analyzed in the next section.

Following the concept, the implemented software tool is subsequently used to determine the performance of the process chain based on the specified parameters and assumptions. Additional values used for the analysis such as CO_2-equivalents and costs are presented in the appendix. The results for alternative process chain configurations are presented in Fig. 5.15.

With regard to the results from the previous case study, the first process chain configuration considers a dry turning process as well as a milling and grinding process using a mineral oil based emulsion. It is assumed that decentral peripheral subsystems

Workpiece material:	C45E, 1.1191	Process:	Slot milling
CF Strategy:	Flooding with 7.5% emulsion and oil, dry machining	Tool:	Solid carbide, ⌀ 10 mm
		Removed volume:	V_w = 5760 mm³
Process:	Longitudinal turning	Process:	Internal cylindrical grinding
Tool:	CCMT 09T304	Grinding wheel:	CBN [B126 M8 VD 49VV 80]
Removed volume:	V_w = 50265 cm³	Removed volume:	V_w = 100 mm³

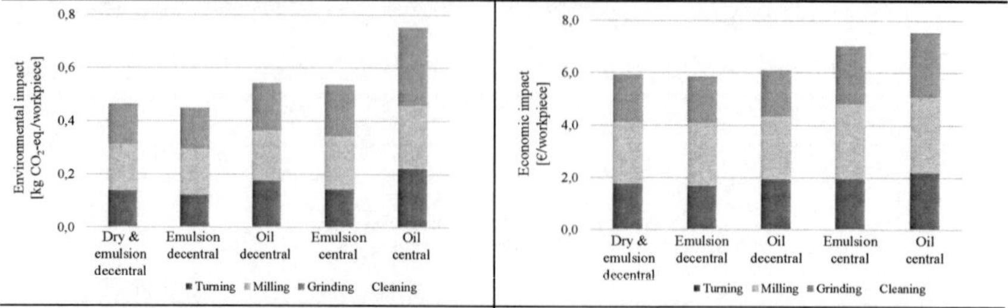

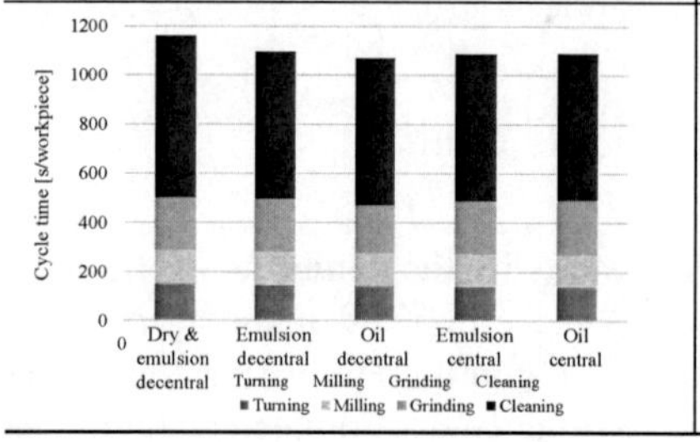

	Dry & emulsion decentral	Emulsion decentral	Oil decentral	Emulsion central	Oil central
Turning					
Cutting speed v_c [m/min]:	163.5	165.2	169.4	185.22	185.9
Feed rate f [mm/rev]:	0.19	0.20	0.20	0.20	0.21
Cutting depth a_e [mm]:	1.50	1.50	1.50	1.50	1.50
Milling					
Cutting speed v_c [m/min]:	85.18	85.13	87.10	89.18	92.14
Feed rate f_z [mm/rev]:	0.04	0.04	0.04	0.04	0.04
Cutting depth a_e [mm]:	1.50	1.50	1.50	1.50	1.50
Grinding					
Cutting speed v_c [m/s]:	89.99	90.0	89.94	90	71.89
Workpiece speed v_w [m/s]:	0.63	0.63	0.75	0.63	0.60
Cutting depth a_e [μm]:	1.50	1.49	1.50	1.49	1.49

Fig. 5.15 Performance of alternative process chain configurations

are used, which also use the same emulsion and do not require a cleaning process in between milling and grinding. In order to remove the scrap material and residuals from the workpiece after dry turning, surface air cleaners are used. To convert the compressed air demand, an electricity input of 0.015 Wh/L is taken into account. The other configurations consider the use of a mineral oil based emulsion or oil for all three machining processes within decentral or alternately central peripheral subsystems. For these configurations, it is again assumed that the same fluid is used and that intermediate cleaning processes are not required, whereas a final cleaning step at the end of the process chain is always considered. In dependence of the configuration, the process parameters are also adjusted by using the developed software tool to minimize the environmental and economic impact as well as the cycle time over the entire process chain, which are equally weighted. The technological impact is again taken into account as a constraint regarding the workpiece surface roughness. This leads to slightly different process parameters for each process chain configuration.

The results reveal that under these assumptions and for the specific process chain, using centralized subsystems especially with oil lead to the highest environmental and economic impacts. This is mainly caused by the energy demand of the cutting fluid system due to an increased pump power and cooling demand. The piping system and thus pressure losses and pump head are significantly increased as well as the required pump power to circulate the cutting fluid within the tank. This also leads to more heat input into the fluid. The increased energy demand, but especially also the higher machine usage rate lead to the increased economic impact. The advantages of centralized cutting fluid systems, which are described by Joksch (2012) in terms of increased process safety and service life of the cutting fluid as well as less maintenance effort are not considered in the context of the analysis.

Also, in decentralized subsystems, the use of oil demonstrates a comparatively higher impact than emulsion, but also a shorter cycle time. By using oil, it is possible to select more demanding parameters for the machining process, but due to the impact of the fluid itself and also higher cooling demands the environmental and economic impacts are also increased. Although the previous investigation demonstrated that dry machining is advantageous for the individual turning system, this is not confirmed within the process chain. In comparison to the use of emulsion with a decentral system it demonstrates slightly higher environmental and economic impacts and also the highest cycle time. This is caused by the required process parameters and also the need for the additional surface air cleaner, which has a high consumption of compressed air and further increases the total cycle time.

5.2.2 Application of Further Improvement Measures

As the case study considers the greenfield planning of a new machining process chain, it is not the target to improve an existing system, but to examine different strategies to achieve the best possible performance. In addition to the alternative configurations and process parameters that were already initially analyzed, the application of selected measures on the use case is presented in more detail in the following.

Standby strategy

The strategy to switch peripheral subsystems off during the standby state counts among the organizational measures as it aims to reduce the time a component consumes power. As described in Sect. 4.4, this measure has already been investigated and implemented for various applications in the context of machine tools and systems. For this case study, the measure is applied to the exhaust air system and filtration/circulation pump. Regarding the exhaust air system, it is important to consider the ramp up time, which may take some seconds and regarding the pump it is important to ensure sufficient circulation to avoid bacterial growth. This measure is generally associated with a low implementation effort and only impacts the environmental and economic, but not the technological performance or the cycle time for the workpiece. The measure is therefore applied to the previously presented process chain configurations and parameters. The effect of this measure is demonstrated with regard to the environmental impact in Fig. 5.16.

The results show a reduced environmental impact for each configuration, which is most obvious for the process chains using centralized peripheral subsystems. These have the highest potential as the consumption of energy due to the pump and cooling unit is particularly high. By reducing the pump utilization, the heat input is also reduced. This measure, however, demonstrates the least impact for the first process

Workpiece material:	C45E, 1.1191	Process:	Slot milling
CF Strategy:	Flooding with 7.5% emulsion and oil, dry machining	Tool:	Solid carbide, ⌀ 10 mm
		Removed volume:	$V_w = 5760\ mm^3$
Process:	Longitudinal turning	Process:	Internal cylindrical grinding
Tool:	CCMT 09T304	Grinding wheel:	CBN [B126 M8 VD 49VV 80]
Removed volume:	$V_w = 50265\ cm^3$	Removed volume:	$V_w = 100\ mm^3$

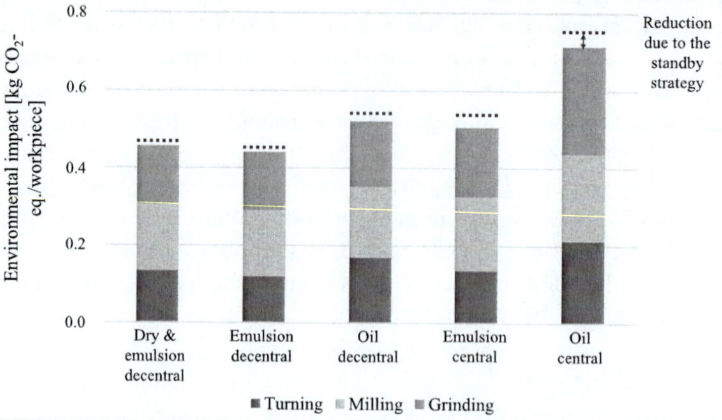

Fig. 5.16 Impact of standby strategy on the environmental impact of the process chain

chain configuration, as it only influences the energy demand of the exhaust air system in the case of dry turning. The economic impact is only slightly influenced by this measure as the labor and machine usage cost rates are the main cost drivers and these are not impacted by it.

In general, the measure is simply implemented in the control unit of a machine tool, but it is important to ensure the cutting fluid quality and operational readiness of the exhaust air system when needed.

Efficient components

Besides the organizational measure, the impact of a technical measure is also examined for the case study. This measure concerns the replacement of specific components by more efficient ones. As previously described the pump motor is such a component due to the high energy consumption of the pump. The planning of a new process chain facilitates the consideration of this strategy, since the implementation does not require any costly retrofitting.

For this case study, the replacement of a motor with the energy efficiency class of IE1 against a motor with the energy efficiency class of IE4 is analyzed. The motors of the supply and filtration/circulation pumps are both considered and again applied to the previous process chain configurations and parameters. The influence on the environmental impact is shown in Fig. 5.17.

The results demonstrate a reduction in impact for each process chain configuration due to this measure. However, in this case no significant difference between the utilization of a centralized compared to a decentralized cutting fluid system could be

Workpiece material:	C45E, 1.1191	Process:	Slot milling
CF Strategy:	Flooding with 7.5% emulsion and oil, dry machining	Tool:	Solid carbide, ⌀ 10 mm
		Removed volume:	$V_w = 5760\ mm^3$
Process:	Longitudinal turning	Process:	Internal cylindrical grinding
Tool:	CCMT 09T304	Grinding wheel:	CBN [B126 M8 VD 49VV 80]
Removed volume:	$V_w = 50265\ cm^3$	Removed volume:	$V_w = 100\ mm^3$

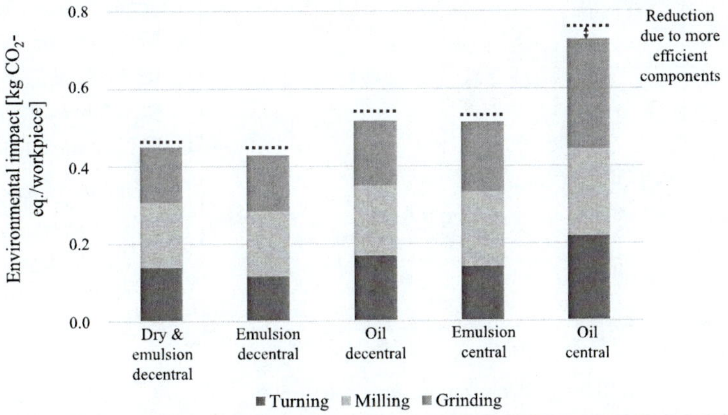

Fig. 5.17 Impact of using a more efficient pump motor on the environmental impact of the process chain

noted. The choice of energy efficiency class influences the resulting motor efficiency and is determined according to Eq. 4.52. Consequently, the influence of different efficiency classes on the motor efficiency is reduced with increasing mechanical pump power. Although centralized cutting fluid systems have a higher pump power and therefore a higher potential, the increase in motor efficiency through this measure is comparatively lower than for decentralized systems. It can further be noted that the impact is slightly higher for process chains using oil instead of emulsion, which is again caused by the reduced heat input and thus required cooling demand. The economic impact is again not significantly impacted by this measure as the labor and machine usage cost are not influenced.

5.3 Overview and Comparison of Improvement Measures

Besides supporting the performance analysis of existing machining systems and the planning of new ones, the concept is also intended to examine the impact of possible improvement measures. Some of these measures have been applied to the previously presented case studies. For a consistent overview, the influence of each measure is shown in Fig. 5.18 in relation to the subsystem in which it is implemented. It is distinguished between the relative potential to reduce the environmental as well as the economic impact for the machining process, cutting fluid and exhaust air system.

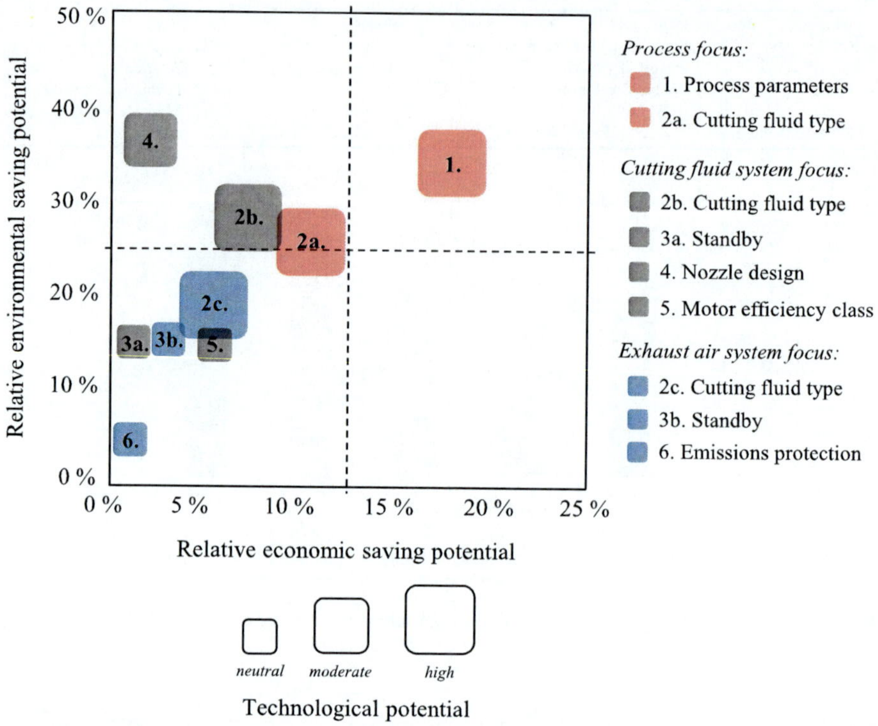

Fig. 5.18 Summary of the impact of the different improvement measures

In view of the concept, the technological impact is not presented with regards to the specific subsystem, as it always focuses on the machining process and workpiece quality. For this overview, the system settings are assumed as previously described and the results are generalized for the different machining processes.

In terms of the machining process, the use of alternative process parameters is evaluated for the cutting process. Due to the possibility to significantly reduce the process time, this measure shows the highest economic and environmental influence. The measure of using different cutting fluids influences all three subsystems and leads to relatively high impact reductions. Both measures are also associated with a high technological potential. Next, the standby strategy can be applied for the cutting fluid and exhaust air system. It is the explicit aim not to influence the technological performance with this measure. In relation to the initial impact of these systems, this measure leads to a significant reduction for the environmental impact, but rather low for the economic impact. It only affects the energy consumption, which is not among the main drivers. This also applied for the measure of using adapted cutting fluid nozzles, but this measure also impacts the technological performance. Its classification, however, only refers to surface grinding, as the nozzles are particularly important for this process. Similarly, using pump motors with a higher efficiency class also leads to a considerably reduced energy consumption of the cutting fluid system, but simultaneously only to slightly reduced costs without impacting the technological performance. The same is valid for the measure of emission protection. By reducing the number of particles that need to be removed it is possible to reduce the energy consumption of the exhaust air system. The saving potential is rather low, but this measure also contributes to the occupational health and safety.

References

Joksch S (2012) Circulation systems for metalworking fluids. In: Astakhov VP, Joksch S (eds) Metalworking fluids for cutting and grinding: Fundamentals and recent advances. Woodhead Publishing, Cambridge, UK, Philadephia, USA

Madanchi N, Winter M, Thiede S, Herrmann C (2017) Energy efficient cutting fluid supply: the impact of nozzle design. In: 24th CIRP international conference on life cycle engineering, Kamakura, Japan, vol 61, pp 564–569. https://doi.org/10.1016/j.procir.2016.11.192

Madanchi N, Leiden A, Winter M, Asbach C, Lindermann J, Herrmann C, Thiede S (2019) Cutting fluid emissions in grinding processes: influence of process parameters on particle size and mass concentration. Int J Adv Manuf Technol 101:773–783. https://doi.org/10.1007/s00170-018-2934-5

VDI 3321 (1994) Schnittwertoptimierung Grundlagen und Anwendung. Verein Deutscher Ingenieure, Berlin, Germany

Chapter 6
Summary, Discussion and Outlook

The final chapter recapitulates the work and summarizes the findings (Sect. 6.1). Following this, the proposed concept is critically discussed (Sect. 6.2) and closes with an outlook by suggesting a future research path (Sect. 6.3).

6.1 Summary

Machining systems are widely used to produce workpieces in large quantities and with complex geometrical shapes. These systems, however, are also relevant in terms of energy and resource consumption, which is strongly connected to the choice of cutting fluid strategy. Cutting fluids are usually used to ensure process stability and high workpiece quality, but influence the need for additional subsystems. In order to foster a sustainable development, the efficiency strategy can be applied by adapting the relation between used input flows and achieved output flows. With regard to machining systems, this could be applied to the consumption of energy, tools or cutting fluid as input flows and the number of produced workpieces as output flows. In view of this situation and challenges, Chap. 1 stresses the significance of cutting fluid strategies on machining systems from an environmental, economic and technological perspective.

Chapter 2 describes the theoretical basis of the research subject. First, the subdivision of machining systems into subsystems and components is defined. The relevant interactions between the subsystems are described by input and output flows. In general, the machining process is considered as the central subsystem, whereas the others are regarded as required support. The connecting element between the subsystems is the cutting fluid for this work. As cutting fluid strategies can be differentiated regarding fluid type and composition, the functions and characteristics of oil- and water-based cutting fluids are discussed. Further, the applied amount of cutting fluid

N. Madanchi, *Model Based Approach for Energy and Resource Efficient Machining Systems*, Sustainable Production, Life Cycle Engineering and Management,
https://doi.org/10.1007/978-3-030-87540-4_6

can also be distinguished and thus MQL and dry machining are also analyzed as alternative strategies. It is concluded that not only the impact on the machining process, but on all subsystems of the machining system including the cycle time must be considered. This leads to an integrated approach, where the impact of each strategy is always use case dependent. In this context, subsystems and components that are influenced by the choice of cutting fluid strategy are further presented in detail. Due to the complex interactions between the subsystems the need for a modeling approach is presented to generate simplified reproductions of the system. These enable the performance of simulation experiments to evaluate the effect of different process settings and alternative cutting fluid strategies. Regarding such evaluations the chapter also describes performance indicators from an environmental, economic and technological perspective. It closes with an overview of preliminary conclusions, which underline the need for a comprehensive approach for the planning and operation of energy and resource efficient machining systems.

Chapter 3 focuses on the state of research by identifying and analyzing research approaches that are within the scope of the work. Due to the large variety of different studies dealing with sustainability of machining systems, the first step defines selection criteria to limit the number to 70 relevant research approaches. They are visualized in a citation network that demonstrates the year of publication, number of citations and information flows between them. The approaches are further classified into the three focus levels of process level, peripheral subsystem level and machining system level. Based on the classification they are further limited to 17 studies of particular interest for the work. In order to analyze and compare them in the context of the research topic, the next step defines relevant and independent criteria for it. Those criteria comprehensively cover the four core areas in terms of research scope, research method and research objective differentiating between the machining process and peripheral subsystem. Based on the structured analysis and comparative assessment of the selected approaches, the research demand is derived.

Chapter 4 demonstrates the developed concept for a technological, economic and environmental design of machining systems by considering the impact of cutting fluid strategies. Based on the analyzed state of research and following the examined core areas, the chapter initially specifies the objectives and requirements. In view of them, the concept is composed of four modules. The first module embraces the system definition with a focus on the energy and resource flows and interactions between the relevant subsystems of a machining system. The modeling of the single machining system elements with the preferred use of semi-empirical models as well as combinations within a process chain are presented within the second module. Based on the modeled energy and resource flows resulting from different process parameters and cutting fluid strategies, the third module focuses on the environmental, economic and technological evaluation of machining systems. For the improvement of the respective machining systems, various organizational and technical measures are presented in the fourth module. The considered measures further differ with respect to the improving object, improving potential and implementation effort. To facilitate the concept application, the models are implemented in an integrated software tool, which automatically determines and evaluates performance indicators for the respective machining system.

Chapter 5 presents the exemplary application of the developed concept to underline its applicability and potential. The application is carried out in the context of two case studies and also indicates the corresponding effects of improvement measures. The performance of three single machining systems is independently analyzed for the first part of the concept application. As part of the use case, the impact of changing parameters and cutting fluid strategies is also demonstrated for each system and considered as a measure to improve the performance. As further technical measure on the peripheral level, the use of an adapted nozzle design to improve the fluid supply as well as the use of a bell-shaped fluid film for emission reduction are examined. The second part of the concept application considers the greenfield planning of a new machining process chain. It analyses the impact of alternative cutting fluid strategies as well as centralized and decentralizes peripheral subsystems. In order to improve the performance of the planned process chain, the organizational measure of switching components off during standby state as well as the technical measure of replacing components by more efficient ones are evaluated. The general impact of the considered improvement measures are summarized at the end of the chapter.

6.2 Discussion

In the context of this work, a concept is developed that aims at the planning and assessment of machining systems not only with regard to technical functions, but also under environmental and economic aspects. The concept has been derived from the requirements and objectives that result from the identified research demand. To meet those, the concept provides a system definition of interactions between machining systems due to the use of cutting fluids, the modeling of the single machining system elements and combinations within process chains, their comprehensive evaluation as well as decision support including exemplary improvement measures.

In comparison with existing research approaches focusing on the assessment or modeling of sustainable machining systems, the developed concept is unique in its approach to comprehensively consider machining systems and their relevant subsystems with regard to the use of cutting fluids. In that context, alternative combinations of process parameters and cutting fluid strategies can be assessed from an environmental, economic and technological perspective. This provides decision support for choosing the most suitable combination of strategy and parameters while planning or operating machining systems. The impact of additional technical and organization improvement measures can also be evaluated. By modeling the relevant energy and resource flows between the elements of the machining system and linking them to GWP, costs as well as workpiece and process quality (e.g. surface roughness, cutting power or tool wear), performance indicators are determined for all three perspectives. Another key feature of the concept, is its modular structure and high degree of flexibility that allows it to be used for various configurations of machining systems with a broad range of applications. The most common machining processes of milling, turning and grinding are considered as well as the relevant subsystems with regard to the cutting fluid strategies of flooding and dry machining. Further, the modular structure allows the approach to be extended to include additional machining processes and strategies. Compared to various other approaches, which require extensive experiments or calculations, the relatively low effort required to apply the concept and its models, is another benefit. As far as possible, semi-empirical models have been employed for which the empirical coefficients are generally available in literature. Their implementation within a software tool further facilitates the usage and ensures a simple application.

These achievements of the proposed concept and modeling approach are contrasted by some aspects that should be viewed critically. First of all, it must be noted that all interactions are statically modeled. Consequently, time-dependent processes cannot be simulated, which may lead to inaccuracies in the calculation of energy and resource requirements. In this context, dynamic simulations could, for example, be relevant for the cooling and cleaning unit as changes in temperature are dependent on time. However, if the system is continuously operated the influence of time-dependent heating is significantly reduced. Furthermore, the accuracy and level of detail of the used models has to be stressed. In accordance to the presented requirements the concept aims at using simplified semi-empirical models with a level of accuracy showing a deviation up to 30%. In most cases this requirement could be satisfied. The drag-out of cutting fluids, however, is modeled using empirical regression models, which show a high level of accuracy, but need to be adapted to the used cutting fluids and respective conditions. Moreover, the grinding process demonstrates a greater deviation compared to the other machining processes. However, a compromise was made to achieve a consistent modeling approach with comparable detail level and effort. While most central models are experimentally validated within the work, an experimental investigation could not be performed for some others such as the centrifuge or cleaning unit due to the lack of accessibility. Although the general concept is applicable to a variety of different strategies, the current models are focused on flooding with different amounts and types of cutting fluid as well as dry machining. The consideration of other strategies might therefore require the addition of models and especially the determination of further coefficients for the existing models, which again have to be validated experimentally.

Despite the existing challenges, which could be addressed to further increase the level of accuracy and extend the applications, the concept application already demonstrated its potential. With little effort it is possible to generally analyze existing or new machining systems as well as to determine and to compare alternative process parameters in combination with cutting fluid strategies.

6.3 Outlook

As already indicated, further research activities on this topic can still expand the concept and contribute to the successful implementation of energy and resource efficient machining systems. An outlook on future research possibilities is briefly demonstrated for the core areas in the following.

Research Scope and Framework

- **Extending the research scope by the factory level** to expand the system boundary and consider further interactions. Currently, the machining system and their relevant subsystems including workpiece and scrap treatment are considered as single machining systems or in combination within a process chain. To further expand the scope relevant subsystems that are influenced at factory level could also be included such as heating, ventilation and air conditioning (HVAC).
- **Integrating the consistency strategy** as an additional strategy to foster sustainable development. The proposed concept focuses on the efficiency strategy as the relationship between input flows to achieve output flows. This strategy is often preferred in practice because it combines ecological and economic benefits, but it also has a short-term focus and builds on existing paradigms. The consistency strategy, in contrast, is designed for the long term, but is also associated with greater effort due to the goal of adapting material and energy flows to fit natural processes. In view of the concept, the application of mineral oil-free cutting fluids based on e.g. vegetable oil (Winter et al. 2012, 2013) can simply be integrated, but requires further knowledge for the adjustment of the coefficients.

Research Object

- **Including additional machining processes** into the proposed concept to enhance the field of application. Due to the high relevance and common use, the concept focuses on milling, turning and grinding. Besides these, other machining processes such as sawing, broaching and honing are also influenced by the choice of cutting fluid strategy. Although they may not be as common, they are also used to fulfill special requirements.
- **Extending the concept by additional cutting fluid strategies** to increase the variety of strategies that can alternatively be applied and compared with each other. While currently only flooding and dry machining are being considered, the modular structure of the concept provides for further strategies such as MQL or supercritical CO_2 (Supekar et al. 2012) to be integrated.

Research Method

- **Employing dynamic simulation methods** could increase the accuracy of determining time-dependent energy and resource flows. With regard to subsystems such as the cooling unit it might be advantageous to couple the proposed concept with a dynamic simulation model. The considerable additional effort for it should be compared to the possible benefit for the respective application.

- **Integrating additional optimization algorithms** into the concept could improve the selection of process parameters in combination with the considered machining system and cutting fluid strategy. Further, the problem-solving time could be reduced for multi-objective improvements. Although the solving time is not an issue at the moment, it could become important in the future, e.g. when coupling with other models and systems.

References

Supekar SD, Clarens AF, Stephenson DA, Skerlos SJ (2012) Performance of supercritical carbon dioxide sprays as coolants and lubricants in representative metalworking operations. J. Mater. Process. Technol. 212:2652–2658. https://doi.org/10.1016/j.jmatprotec.2012.07.020

Winter M, Bock R, Herrmann C (2013) Investigation of a new polymer-water based cutting fluid to substitute mineral oil based fluids in grinding processes. CIRP J. Manuf. Sci. Technol. 6:254–262. https://doi.org/10.1016/j.cirpj.2013.07.003

M. Winter, T. Öhlschläger, T. Dettmer, S. Ibbotson, S. Kara, C. Herrmann, Using Jatropha oil based metalworking fluids in machining processes: a functional and ecological life cycle evaluation. in *19th CIRP International Conference on Life Cycle Engineering* (Berkeley, USA, 2012), pp. 311–317. https://doi.org/10.1007/978-3-642-29069-5_53

Appendix

Regression Parameters

Regression coefficients to determine avergage exhaust air flow (Sect. 4.2.3.2)

Machining with undefined cutting edges	Oil	Emulsion	Dry
a_{ea}	490	230	743
b_{ea}	372	365	360
Machining with defined cutting edges			
a_{ea}	215	105	352
b_{ea}	309	285	325

Regression coefficients drag-out via emissions for grinding (Sect. 4.2.3.2)

Coefficient	Oil	Emulsion
$K_{cf,ea}$	−4.10E−02	3.85E−03
$C_{ea,g}$	3.61E−04	1.42E−04
α^1	1.54E−01	8.50E−01
α^2	9.74E−01	1.81E−01
α^3	1.68E−01	−1.85E−01
α^4	7.88E−01	4.75E+01
α^5	−1.03E−01	6.51E−01

N. Madanchi, *Model Based Approach for Energy and Resource Efficient Machining Systems*, Sustainable Production, Life Cycle Engineering and Management,
https://doi.org/10.1007/978-3-030-87540-4

Regression coefficients drag-out via emissions for turning (Sect. 4.2.3.2)

Coefficient	Oil	Emulsion
$K_{cf,ea}$	2.85E−04	−1.97E+00
β^1	4.12E−01	1.41E−02
β^2	9.11E−01	5.24E−02
β^3	9.62E−02	2.09E−02
β^4	−6.26E−01	−2.49E−02
β^5	4.82E−01	8.47E−02

Regression coefficients drag-out via workpiece wetting after turning (Sect. 4.2.3.3)

Coefficient	Oil	Emulsion
a_{ws}	3.65E+00	7.47E−01
b_{ws}	1.20E+01	2.33E+00
c_{ws}	−3.58E−01	−6.45E−02

Regression coefficients drag-out via workpiece wetting after milling (Sect. 4.2.3.3)

Coefficient	Oil	Emulsion
a_{ws}	4.23E+00	1.23E+00
b_{ws}	1.14E+01	4.07E+00
c_{ws}	−3.73E−01	−7.10E−02

Regression coefficients drag-out via chips wetting after turning and milling (Sect. 4.2.3.3)

Coefficient	Oil	Emulsion
a_{ws}	2.48E−01	2.13E−01
b_{ws}	2.94E−01	2.02E−01
c_{ws}	−1.49E−03	−2.16E−03

Regression coefficients drag-out via swarf wetting after grinding (Sect. 4.2.3.3)

Coefficient	Oil	Emulsion
a_{ws}	1.24E+00	1.08E+00
b_{ws}	−1.92E+00	7.47E−01
c_{ws}	−2.84E−02	−2.96E−03

Case Study Specification

Specification of considered input parameters (Chap. 5)

	Parameter	Symbol	Unit	Value
Cutting force specification - workpiece	Specific force	$k_{c1.1}$	[N/mm^2]	2220
	Constant for cutting force	mc	[–]	0.14
Tool specification - turning	Corner radius of the cutting tool	r_ε	[mm]	0.4
	Rake angle of cutting tool	γ	[°]	7
Cutting force specification - turning	Correction factor CF strategy - oil	K_{cf}	[–]	0.85
	Correction factor CF strategy - emulsion	K_{cf}	[–]	0.9
	Correction factor for CF strategy - dry	K_{cf}	[–]	1.0
	Correction factor for cutting materials	K_m	[–]	1.0
	Correction factor for cutting process	K_{vk}	[–]	1.0
	Correction factor for tool wear	K_{tw}	[–]	1.0
Tool wear specification - turning	Constant tool wear	k	[–]	−4.01
	Constant tool wear	k_a	[–]	0.00
	Constant tool wear	k_f	[–]	−0.78
	Constant tool wear	C	[–]	6.91E+09
	Correction factor CF strategy - flooding	K_{cf}	[–]	1.0
	Correction factor CF strategy - dry	K_{cf}	[–]	0.8
Tool specification - milling	Milling tool diameter	D	[mm]	10.0
	Depth of cut	a_e	[mm]	10.0
	Number of cutting edges	Z	[–]	4
	Corner chamfer width at 45°	r_ε	[mm]	0.1
	Cutting direction angle	φ_c	[°]	180
	Rake angle of cutting tool	γ	[°]	30
Cutting force specification - milling	Correction factor CF strategy - oil	K_{cf}	[–]	0.85

(continued)

(continued)

	Parameter	Symbol	Unit	Value
	Correction factor CF strategy - emulsion	K_{cf}	[–]	0.9
	Correction factor for CF strategy - dry	K_{cf}	[–]	1.0
	Correction factor for cutting materials	K_m	[–]	1.0
	Correction factor for cutting process	K_{vk}	[–]	1.15
	Correction factor for tool wear	K_{tw}	[–]	1.0
Tool wear specification - milling	Constant tool wear	k	[–]	−1.98
	Constant tool wear	k_a	[–]	−0.21
	Constant tool wear	k_{ae}	[–]	−0.12
	Constant tool wear	k_f	[–]	−0.56
	Constant tool wear	k_z	[–]	−0.68
	Constant tool wear	C	[–]	3.86E+04
	Correction factor CF strategy - flooding	K_{cf}	[–]	1.0
	Correction factor CF strategy - dry	K_{cf}	[–]	0.8
Tool specification - grinding	Effective grain spacing	λ_{ke}	[mm]	45
	Diameter of the grinding wheel	d_g	[mm]	40
	Width of cut	a_p	[mm]	10
	Grain tip radius	r_g	[mm]	0.013
	Dressing depth	a_{ed}	[mm]	0.001
	Hardness of the grain surface	$H_{s,g}$	[MPa]	56,000
	Hardness of the bulk material of the grain	$H_{b,g}$	[MPa]	45,000
	Usable grinding wheel volume	V_{gw}	[mm^3]	21,991
Cutting force and tool wear specification - grinding	Correction factor CF strategy - emulsion	K_{cf}	[–]	1.0
	Correction factor for CF strategy - oil	K_{cf}	[–]	0.9
	Dimensionless distance between adjacent tracks	l	[–]	0.2
	Interfacial friction coefficient - oil	f_{cf}	[–]	0.32

(continued)

(continued)

	Parameter	Symbol	Unit	Value
	Interfacial friction coefficient - emulsion	f_{cf}	[–]	0.28
Cutting fluid specification - Mineral oil based emulsion	Density of cutting fluid	ρ_{cf}	[kg/m^3]	1010.0
	Kinematic viscosity of cutting fluid	ν_{cf}	[mm^2/s]	0.87
	Surface tension of the cutting fluid	σ_{cf}	[mN/m]	31.17
	Specific heat capacity of cutting fluid	$c_{p,cf}$	[kJ/(kg·K)]	3.342
Cutting fluid specification - Mineral oil	Density of cutting fluid	ρ_{cf}	[kg/m^3]	890.0
	Kinematic viscosity of cutting fluid	ν_{cf}	[mm^2/s]	12.0
	Surface tension of the cutting fluid	σ_{cf}	[mN/m]	29.65
	Specific heat capacity of cutting fluid	$c_{p,cf}$	[kJ/(kg·K)]	2.094
Average cutting fluid system specification - varies within case study	Cutting fluid volume flow	$\dot{V}_{cf}$	[l/min]	25
	Total cutting fluid volume	$V_{cf,total}$	[l]	500
	Outlet area of the nozzle	A_n	[m^2]	0.000016
	Flow velocity	v	[mm/s]	2500
	Pump head	H_P	[m]	1.15
	Ambient air Temperature	T_1	[°C]	21
	Cutting fluid component temperature	T_2	[°C]	25
	Energy efficiency ratio	EER	[–]	0.9
	Emission level of steel surface	ε_1	[–]	0.8
Average exhaust air system specification - varies within case study	Total pressure increase	Δp_V	[bar]	1.0
	Exhaust air fan efficiency	η_{ea}	[–]	0.85
	Size of the workstation volume	V_{ws}	[m^3]	2.0
Average cleaning unit specification	Density of cleaning fluid	ρ_{cm}	[kg/m^3]	990.0
	Specific heat capacity of cutting fluid	$c_{p,cm}$	[kJ/(kg·K)]	4.168
	Volume of cleaning medium	V_{cm}	[l]	100

(continued)

(continued)

	Parameter	Symbol	Unit	Value
	Cleaning time for workpieces	t_{cm}	[s]	600
	Time to remove workpieces from cleaning unit	$t_{cm,out}$	[s]	10
	Dripping time	t_{ws}	[s]	5
	Ambient air Temperature	T_1	[°C]	21
	Desired cleaning fluid temperature	T_2	[°C]	70
	Energy efficiency ratio	EER	[–]	0.9
	Fluid transfer coefficient	β_m	[m/h]	7.0
	Evaporation heat at 70 °C	r	[kj/kg]	2333.10
	Specific gas constant for water vapour	R_D	[J/(kg·K)]	46152
Average de-oiling unit specification	Density of workpiece	ρ_w	[g/mm^3]	0.00785
	Dripping time	t_{ws}	[s]	20
	Acceleration time for de-oiling	$t_{de,a}$	[s]	5
	Capacity of centrifuge for de-oiling	m_{de}	[kg/min]	2
	Thickness of the centrifuge	s	[m]	0.025
	Centrifuge height	H_c	[m]	0.5
Equivalent CO_2 impact for environmental performance	due to produced cutting fluid - oil	$CO_{2,cf,p}$	[kg-eq/kg]	0.49
	due to disposed cutting fluid - oil	$CO_{2,cf,d}$	[kg-eq/kg]	1.32
	due to produced cutting fluid - emulsion	$CO_{2,cf,p}$	[kg-eq/kg]	0.02
	due to disposed cutting fluid - emulsion	$CO_{2,cf,d}$	[kg-eq/kg]	0.14
	due to energy (energy mix Germany)	$CO_{2,e}$	[kg-eq/kwh]	0.534
	due to tools - turning	$CO_{2,t}$	[kg-eq/cutting edge]	0.10
	due to tools - milling	$CO_{2,t}$	[kg-eq/tool]	3.68
	due to tools - grinding	$CO_{2,t}$	[kg-eq/tool]	2.79
Prices for economic performance	Cutting fluid procurement price - oil	$p_{cf,p}$	[€/kg]	2.6

(continued)

(continued)

	Parameter	Symbol	Unit	Value
	Cutting fluid disposal price - oil	$p_{cf,d}$	[€/kg]	0.07
	Cutting fluid procurement price - emulsion	$p_{cf,p}$	[€/kg]	0.23
	Cutting fluid disposal price - emulsion	$p_{cf,d}$	[€/kg]	0.07
	Energy procurement price	p_e	[€/kWh]	0.15
	Tool price - turning	p_t	[€/cutting edge]	7.5
	Tool price - milling	p_t	[€/tool]	67.6
	Tool price - grinding	p_t	[€/tool]	361
	Labor rate	$p_{l,i}$	[€/h]	12.26
	Machine hour rate - turning	$p_{m,i}$	[€/h]	20.78
	Machine hour rate - milling	$p_{m,i}$	[€/h]	22.80
	Machine hour rate - grinding	$p_{m,i}$	[€/h]	15.79